PLANET EARTH

From 4.5 billion years ago to the present

Dedication

To my loving wife Ann – a unique person, wholly at one with the planet and all forms of life, who has travelled the world with me and who has inspired me to write this book – thank you for everything.

First published in November 2015

A catalogue record for this book is available from the British Library

ISBN 978 0 85733 810 5

Library of Congress control no. 2014957832

Published by Haynes Publishing,
Sparkford, Yeovil,
Somerset BA22 7JJ, UK.

Tel: 01963 440635
Int. tel: +44 1963 440635
Website: www.haynes.co.uk

Haynes North America Inc.,
861 Lawrence Drive, Newbury Park,
California 91320, USA.

Printed in the USA by Odcombe Press LP,
1299 Bridgestone Parkway,
La Vergne, TN 37086.

PLANET EARTH

From 4.5 billion years ago to the present

Owners' Workshop Manual

The practical guide to the origin, evolution and future of Earth

David Baker

Contents

OPPOSITE A view of the Grand Canyon from Mohave Point with the Colorado River. *(USGS).*

RIGHT Devoid of life, the early Earth would have been bombarded with material left over from the birth of the solar system. *(David Baker)*

Chapter One

Earth's place in the solar system

To see the Earth as it truly is, small and blue and beautiful in that eternal silence where it floats, is to see ourselves as riders on the Earth together, brothers on that bright loveliness in the eternal cold – brothers who know now they are truly brothers.

— Archibald MacLeish, 1968

OPPOSITE Dramatic vistas, typified by the Grand Canyon in Arizona, seen here from Yavapai observation point, have inspired centuries of travellers. Early explorers were struck by the awesome magnitude of a planet they little understood, but explanations of the world around them had toawait the advent of scientific process. *(Tobias Alt)*

RIGHT A Scottish geologist and physician, James Hutton (1726–1797) espoused the theory of uniformitarianism in which all the physical structures of Earth evolved over time, claiming them to be natural processes and not placed by divine intervention. *(Raeburn)*

FAR RIGHT A prominent geologist of the 18th century, Charles Lyell (1797–1875) helped to unlock the secret of Earth's origin from the constraints of religious dogma through his book *Principles of Geology*, published in three volumes 1830-1833. *(Alexander Craig)*

RIGHT Inspired by the work of Charles Lyell, Charles Darwin (1809-1882) acquired international fame for his work on the evolution of life, challenging the Biblical teaching of the day that all living things had been created in seven days. *(G Richmond)*

Earth is a giant engine, built by natural forces to evolve and produce life, of which we as humans are a component in the machine. Once begun, it is a continually changing structure with a motor in its interior and a very thin gaseous envelope that has itself been the product of life. All the different types of life produced by this engine have taken a significant role in changing its outer surface and its atmosphere so that the motor is sustained by a replicating process, continually manufacturing the fuel it needs to survive, moving it around on the surface and returning it for reprocessing and regurgitation as new fuel for the motive forces that keep the rocks moving.

But within a very thin sliver of time humans have emerged from those various life forms and have begun to interpret the machine, some believing it to have been manufactured by a supreme Being, others believing that, as a natural product of an evolving process, we as humans have the right to use it as we see fit – as a garden, a playground and a larder; that humans have the right to 'manage' Earth and all other life forms and to fashion the planet for human replication and gratuitous consumption as they see fit.

But ideas are evolving fast and many people believe that Earth is not there either as a gift from several competing gods, or from a single all-pervading God, but rather as a platform on which humans can coexist with life forms that have populated the planet for millions of years. It is a human journey in understanding Earth that has, for all the brevity of human existence on this planet, taken a long time to evolve from some very different ideas and interpretations about just what Earth is at all.

Earth was once thought to be at the centre of the universe; then it was displaced to being just one of several bodies floating in orbit around the Sun. Now we no longer believe even our Sun is at the centre of the universe, but is one of many billions of stars in a galaxy we call the Milky Way. Not only that, our Milky Way galaxy is now known to be but one of *billions* of other galaxies in a universe still expanding and getting bigger.

So does that leave our little Earth now a meaningless lump of rock, so far removed from our presumption of its importance that it has nothing special to offer, and we upon it are nothing more than an assemblage of atoms and molecules? Not at all. Astronauts journeying back to Earth from the Moon regarded it as a place so special in the blackness of space that they were moved to believe it had, in humankind's quest for the outermost reaches, brought humans themselves back to a realisation that Earth is indeed special – and a very special place to treasure.

The space age begat an age of environmental concern, for having seen Earth as a blue marble in the velvet blackness of space, humans were moved in a way they had never been touched before. And it changed everything. Concern for Earth and for what our pillaging of limited resources could mean for future generations sparked waves of protest and action, against abuse of the planet and mindless destruction of the environment. A new awareness began to take hold and it changed forever the way we see ourselves and our planet.

This book is about the physical workings of Earth, and to understand how it operates requires first an awareness of where precisely it is in the solar system, what keeps it operating as it does, and what changes it may undergo in the next several centuries and millennia. Only then will it become clear how it works and in what manner it operates on a sustained basis, supporting life and maintaining a comfortable environment.

What we now understand about Earth and our solar system is the product of scientific investigation and interdisciplinary work. Modern scientific methods of investigation involve instruments and equipment that were unknown just a few decades ago. These investigations have dramatically reinterpreted the story of Earth and transcend the limited and simplistic view of the solar system most people had until the second half of the 20th century. Until then it was a simple and enduring arrangement of worlds, comets and asteroids, never changing and fixed in their orbits.

As formerly understood, the solar system was neatly divided into two zones: the inner group of four planets with rocky surfaces, defined as terrestrial planets; and an outer

BELOW William Smith (1769-1839) produced the first geological map of the British Isles in 1815, but his lack of formal education condemned him to obscurity. He spent a period in debtors' prison when the Geological Society of London took his maps and sold them for a cheaper price. *(Alan Levine)*

zone of four giant planets primarily formed of gas around a small rocky core deep at their centre. We now know there is a third zone, an outer region populated by small icy bodies the

largest of which is Pluto, once thought to be one of the main planets in the solar system. These are known as Kuiper Belt bodies, named after the Dutch-American astronomer Gerard Kuiper, but credit should also go to Kenneth Essex Edgeworth (1880–1972) who raised the possibility of large quantities of icy bodies beyond Neptune. We also now know that there is a fourth region, an enveloping cloud of comets and ice stretching at least a quarter the way to the nearest star and this is known as the Oort Cloud.

The terrestrial planets, of which Earth is the largest, are separated from the gaseous outer giants by a disc of small rocky fragments scaled in size from fractions of a millimetre to a few kilometres, but too small to have gravitationally pulled themselves into a spherical shape. This is the asteroid belt, where billions of tiny rocky fragments are distributed in a wide band of space that space scientists once feared would be an obstacle to spacecraft traversing the region on their way to the outer giants. In 1972 and 1973, two spacecraft were launched to fly past Jupiter and to do that they had to fly through the asteroid belt. Both survived, as have the other six spacecraft launched to the outer planets and beyond since that time.

The third region – the Kuiper Belt – has been recognised as a defined part of the solar system only in the last few decades, long after the discovery of Pluto by Clyde Tombaugh in 1930. At first astronomers believed that Pluto was one of the main planets in the solar system, albeit a very small one, until unease about that resulted in the discovery of other dwarf worlds all orbiting the Sun. This resulted in the decision to define Pluto as a 'dwarf planet', the most prominent member of the Kuiper Belt.

The outer region very much farther away, known as the Oort Cloud, is named after the Dutch astronomer Jan Oort. This is believed to contain millions of comets and takes the form of a spherical shell occupying a vast region of space at the very edge of the Sun's gravitational influence. These icy objects represent the primordial material that was around when the Sun first formed, and their very existence is made possible only by their remoteness from the dynamic events which were to form the eight main planets and the dwarf planets of the Kuiper Belt.

To appreciate the scale of the solar system and the distances involved it is best measured in multiples of astronomical units (AU), where one AU equals the distance of Earth from the Sun, on average 93 million miles (150 million km). The two inner planets, Mercury and Venus, are at a distance from the Sun of about 0.3AU and 0.7AU respectively. Mars, outside Earth's orbit of the Sun, lies 1.5AU distant, and the asteroid belt occupies a broad swathe of the solar system between 2.1AU and 3.3AU.

Largest of the four gaseous outer giants, Jupiter is 5.1AU from the Sun, followed by the ringed planet Saturn at 9.5AU, Uranus at 19.2AU and Neptune at 30.1AU. The Kuiper Belt bodies lie beyond Neptune, although Pluto is in an orbit that for 20 years in its 248-year path around the Sun comes inside the mean radius of Neptune's orbit. That occurred last between 1979 and 1999. Not until 2227 will it once again be just a little bit closer to the Sun than the mean radius of Neptune's path. The orbit of Pluto is elliptical, lying between 29.6AU and 48.8AU at its most distant point. Right now Pluto is moving farther away from the

Sun and will reach the farthest point in its path around 2113.

Only on this scale of measurement can the extreme distance of the Oort Cloud be fully comprehended. The cloud itself is believed to be divided into two zones: a doughnut-shaped inner cloud between 2,000AU and 20,000AU; and a spherical enveloping cloud of comets between 20,000AU and about 50,000AU. Put another way, the inner zone of the Oort Cloud is more than 40 times the distance of Pluto at the farthest point in its orbit. It took NASA's New Horizons spacecraft more than nine years to fly from Earth to Pluto. It will take that spacecraft more than 360 years to reach the Oort Cloud.

The entire cloud of comets may account for trillions of objects. When individual members are distorted in their path around the Sun their paths are changed into highly elliptical orbits of the Sun, bringing them down so that the closest point of approach to the Sun is in the zone of the terrestrial planets, less than 5AU from the Sun and frequently very much closer.

The geometry of the solar system was believed by astronomers to be an ordered place, with planets in almost circular orbits, the outermost being Pluto, reaching farther from

LEFT Wegener writes up his notes during an expedition to Greenland in 1912, when expedition leader Johan Peter Koch fell into a crevasse and broke his leg. *(Bundesarchiv)*

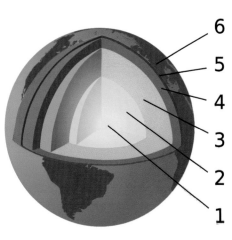

the Sun than any other planet. But apart from Earth only five planets were visible to the ancients. Uranus was discovered in the 18th century, Neptune in the 19th century and Pluto in the 20th century. By the latter decades of that century the solar system was known to be a little different to

LEFT A simplified cutaway diagram of Earth's structural makeup as deciphered by successive generations of geologists and geophysicists. Key: 1 inner core; 2 outer core; 3 lower mantle; 4 upper mantle; 5 & 6 lithosphere, ocean and continental crust.

M E S O Z O I C E R A
251 MILLION YEARS AGO
TRIASSIC PERIOD
JURASSIC PERIOD
CRETACEOUS PERIOD
65 MILLION YEARS AGO
Paleocene Epoch
Eocene Epoch
Oligocene Epoch
Miocene Epoch
Pliocene Epoch
Pleistocene Epoch
Holocene Epoch
TERTIARY PERIOD
QUATERNARY PERIOD
C E N O Z O I C E R A
PERMIAN PERIOD
PENNSYLVANIAN PERIOD
MISSISSIPPIAN PERIOD
DEVONIAN PERIOD
SILURIAN PERIOD
CAMBRIAN PERIOD
ORDOVICIAN PERIOD
P A L E O Z O I C E R A
PRECAMBRIAN
1 BILLION YEARS AGO
2 BILLION YEARS AGO
3 BILLION YEARS AGO EARLIEST ORGANIC STRUCTURES
4.5 BILLION YEARS AGO

the ordered symmetry it was once thought to follow.

The space age has provided unprecedented impetus to planetary research and to the science of the solar system, as well as Earth. Direct measurement of the edge of the solar system by NASA's Voyager spacecraft shows scientists that the Sun's magnetosphere, known as the heliosphere, is dominant out to about 110AU, although that precise distance depends on the strength of the solar wind of charged particles flowing across the solar system. This defines the outer limits of the Sun's heliospheric influence, a distance twice as far from the Sun as Pluto is at its farthest point.

But that is not the outer edge of the Sun's gravitational influence, which extends to at least 50,000AU, where the outer shell of the Oort Cloud resides. Some astronomers believe the cloud extends as far as 100,000AU from the Sun, which would represent a distance halfway to the nearest star, Proxima Centauri. Whichever of the two figures is correct, the edge of the solar system can be expressed as lying between one and two light years distant, which represents the outer edge of the Sun's gravity field and its influence. Beyond that is the region of interstellar space where the Sun has no influence, either from its heliosphere or its gravity.

ABOVE The geological ages of the Earth, illustrated on an imaginative spiral using artistic licence to display the burgeoning spread of life upon a geologically active planet that has never been at rest. Earth will continue to reflect the changes brought to it by living organisms. *(USGS)*

ABOVE Earth has been photographed by many satellites and spacecraft since the 1960s, revealing the planet to be a world dominated by the presence of water and a series of continental plates densely populated with flora and fauna, enveloped by an atmosphere which is itself the product of life. (NASA)

BELOW Along with Venus (second from left), its sister planet, Earth dominates the inner solar system, alongside tiny Mercury (left) and Mars (right). (NASA)

All this has transformed our view of the way the solar system has evolved, and it is now understood to be in a great state of flux, where change is normal and worlds come and go. Direct observation of Earth from its surface, from its oceans, from the air and from space has allowed scientists to build a picture of an evolving planet which has gone through many turbulent ages and is continuing to do so into the future.

One of the more profound changes to understanding Earth came as a result of the exploration of the Moon, with unmanned spacecraft since the mid-1960s and with the manned Apollo missions between 1969 and 1972. The return of lunar samples for laboratory analysis opened new fields of study that provided detailed insight on the early period of the solar system. Where on Earth the first one-third of geological history has been largely eroded by the active changes to its surface through

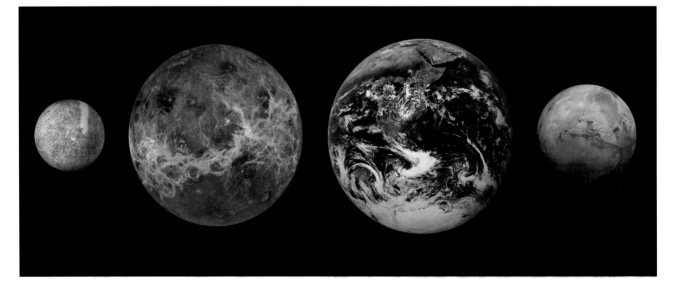

weathering and plate tectonics, the Moon is largely locked in the first third of its history. And because the Moon and Earth formed approximately at the same time, what we are denied on Earth we find on the Moon, as evidence of what the solar system was going through in that very early period of its evolution.

In this way the evolution of Earth and the Moon are tied together, and a comprehensive study of the evolution of the Moon has helped unlock previously unknown aspects of their combined history. The Earth-Moon system, locked together gravitationally and in the combined interaction with Earth's geomagnetic field, is now regarded as a bi-planetary system, in all but name. Detailed analysis of data continually transmitted to Earth for several years from several instruments left at five of the six lunar landing sites has provided evidence of this interaction. Earth would not have evolved along a path sympathetic to the evolution of life – perhaps even its origin on this planet – had it not been for the existence of the Moon, about which more later on.

Other changes to our previous view of an orderly solar system come from detailed computational models of other worlds derived from information returned from satellites and spacecraft, defining the exact position of the planets and their specific sizes with great precision. Using models constructed from observation and measurement, it is possible now to say that the solar system has gone through epic periods of change with great violence and turmoil, and that what we see today is equivalent to observing a single picture frame on a running motion film of extraordinary events that have, for a long time, defied logic.

The planets did not form in the places they occupy today, and all have gone through turbulent times and undergone changes both dramatic and extreme, causing upheaval to atmospheres, climatic conditions and movement across the solar system that at times have threatened their very existence. It has been that way with the story of how we understand the solar system and it is that way with the view scientists now have about Earth itself.

Summary

- ■ Understanding Earth's true place in the universe has involved philosophy, religion and science.
- ■ The solar system is divided into inner and outer zones by an asteroid belt.
- ■ The evolution of Earth and Moon are an integrated story.

Chapter Two

How Earth began

To understand the origin of Earth it is first necessary to understand the origin of our star – the Sun – and the solar system of which we are an integral part. A considerable amount of information has been collected since the 1960s, leading to a dramatic revision in the way scientists understand the origin of the solar system and its planets.

OPPOSITE Very young hot stars formed early in the history of the Universe this cluster, at a distance of 12 billion light years, being typical of the energetic activity taking place within the first few billion years. Star production has slowed dramatically as the material from which thermonuclear reactions begin is consumed. *(NASA)*

Some of that information has come from direct observation using ground-based telescopes and some has come from space-based instruments, while much has been gathered from data transmitted by spacecraft on or around the planets in the solar system – including Earth itself.

A lot of information has also been gleaned about the origin of the solar system by combining the results of studies by solar scientists, astrophysicists and astronomers working with theoretical physicists to more fully interpret what has become a rounded and fully inter-disciplinary subject.

In determining the origin and evolution of Earth itself, all sorts of disciplines are involved, including geophysics, geochemistry, geomorphology and geology. The development and evolution of the atmosphere requires its own atmospheric physicists, climatologists and meteorologists. And when it comes to fathoming the origin of carbon-based life on Earth, chemistry, biology, biochemistry and biophysics all play a role.

For determining the evolution of living things, botanists and palaeontologists are also needed; and when it comes to humans, physical anthropologists and anatomists are added to the mix until archaeologists help define the last several thousand years and explain the way humans have irrevocably changed the way Earth is evolving.

To understand where Earth came from, and to know the structure that forms it today, we need to trace not only the origin of the building blocks of matter but also the forces that prevailed at the time of its formation and continue to exist today. Thus, it is necessary to briefly understand the origin of the solar system and the universe within which the Sun – and all its attendant worlds, physical bodies and gases – formed. Those events began when the universe emerged as a single, unifying event. The story of Earth begins more than eight billion years before it formed as an independent planet in the solar system.

Space and time

Scientists who study the origin and evolution of the universe are called cosmologists. They believe that the universe began almost 13.8 billion (13.8×10^9) years ago in a sudden materialisation known as the Big Bang; but could this also have been the beginning of time? In classical astronomy, time certainly had to be in existence at the point the universe materialised – became matter – so that a sequence of events could occur. Surely, without time there is no sequence against which to measure change?

But the question as to whether time existed *before* the universe is outside the realm of classical scientific description. It is said that if it did exist, then the universe we call '*the*' universe may not have been the first universe, because that presupposes the prior existence of a scale to measure the origin of the universe. Neither might it have been the only universe. And it might not be the only one today in our present understanding of time.

Indeed, multiple universes may have existed

before the Big Bang; our universe may be the product of a violent event in one or more other universes, or we may be coexisting within a matrix of multiple dimensions – much like multiple radio signals come into our laptops and smartphones for communication, navigation and information, without any one signal knowing (or caring) about the existence of the others.

But the existence of multiple universes – either now or before the origin of our universe – is a subject outside the realm of logical explanation in the traditional sense and, although many dissertations have been written on such possibilities, conclusions are

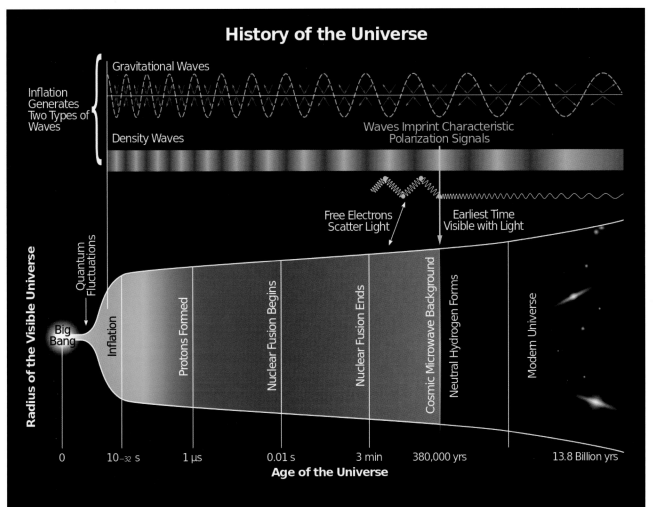

History of the Universe

Inflation Generates Two Types of Waves

Gravitational Waves

Density Waves

Waves Imprint Characteristic Polarization Signals

Free Electrons Scatter Light

Earliest Time Visible with Light

Radius of the Visible Universe

Quantum Fluctuations

Big Bang

Inflation

Protons Formed

Nuclear Fusion Begins

Nuclear Fusion Ends

Cosmic Microwave Background

Neutral Hydrogen Forms

Modern Universe

| 0 | 10^{-32} s | 1 µs | 0.01 s | 3 min | 380,000 yrs | 13.8 Billion yrs |

Age of the Universe

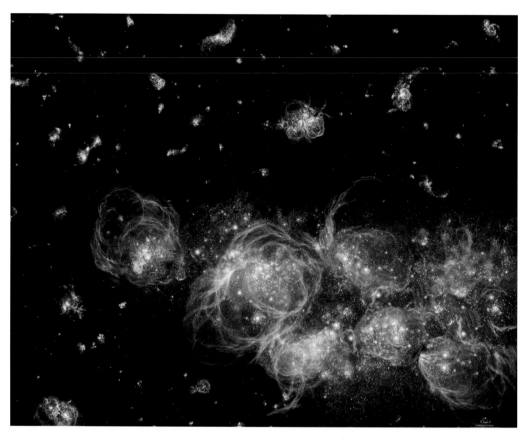

speculative, albeit bound by mathematical plausibility. In fact, we now believe that models relevant to the macroscopic view of the universe at large and the microscopic world of quantum mechanics were tied together to provide a solution which explains why there was no 'before' the universe began.

Time and space were formed at the same point because the existence that we understand requires a four-dimensional world-view: three spatial dimensions (up/down, left/right, fore/aft) and time. The very existence of spatial dimension requires time to set the 'clock' of motion that determines relativity. Laws that exist today in the macroscopic world of general relativity do not work on the very small-scale structure of matter, which is ruled over by the quantum world.

If time is wound back to the beginning of the universe, gravitational forces were so intense they compressed into a single point the macroscopic with the microscopic, and there was no difference. The difference we see and describe today only occurred after the origin of time and the universe began to inflate, taking the two states of law along separate paths.

So, for the purposes of sticking to a rational and scientific model for understanding how Earth works, we must accept that what we perceive to be true today is the point from which we can begin, and the place where we can begin to explain the origin of the materials necessary to build our planet is the point at which the fundamental building blocks came together.

When we say that the universe is 13.8 billion years old, a common misconception has it that it must be no more than 27.6 billion light years across; expansion across time, it is said, would have caused 'space' to have expanded into a spherical volume with a radius of 13.8 billion light years. But the universe has not stood still in the 13.8 billion years since light from the most distant source reached us. The distance today from any central observation point we look from (in our case, Earth) to the edge of the spherical universe is almost 47 billion light years, a diameter of up to 94 billion light years.

Because the observable universe is a sphere centred on the observer, it has an equal radius from that position whether viewed from this position on Earth or another position billions of light years away. From wherever the 'observer' is positioned, the universe appears uniform and of equal radius, always (approximately) 13.8 billion years old.

In the beginning...

When the universe began at the start of time, it was very small, dense and devoid of matter, which we define as structures composed of atoms – indeed, atoms did not even exist. It was from the singularity that a rapid expansion occurred which would eventually lead to the stars, galaxies and matter that we observe today.

There are several nuanced descriptions for the very early period in the history of the universe, but the earliest known point is referred to as Planck time, defined as the smallest meaningful description of a length to time. But this is equally uncertain – hence, it emerges mathematically from Heisenberg's Uncertainty Principle. This states that a particle cannot be precisely tied down to a defined location at a specific time and that it cannot be predicted with total accuracy where it will be at a defined point in the future.

Planck time is defined as the time a photon would take to travel, at the speed of light, a distance known as the Planck length, a distance where quantum effects dominate and where the normal functions of gravity and time appear to cease. It is expressed as 1.616×10^{-35}m (636.30×10^{-36}in), equal to 10^{20} times the size of a proton. Planck time is calculated, theoretically, to be about 10^{-44} seconds, and scientists have directly measured down to a duration of 10^{-17} seconds. This is when time emerged as a dimension.

All this is important because it is from this start point that we explain the distribution of elements from which our solar system – and Earth – formed, and this Planck epoch lasted probably no longer than 10^{-32} seconds. In this period, temperatures were phenomenally high and only one force existed in this universe, which was incredibly dense and minute. This single force in fact combined the four forces (not dimensions) which would eventually separate out and from which actions and reactions would produce all the matter on which our universe, and Earth, runs today.

Following this the universe entered the grand unification epoch, beginning about 10^{-43} seconds after the Big Bang, where temperatures had cooled to about 10^{30}K (the Kelvin scale establishes zero as absolute zero where all molecular movement stops; the size of one unit of Kelvin equals one degree Celsius) and where gravity emerged from the single force leaving what is called the electronuclear force. As the universe cooled it crossed phase transition barriers. This phase lasted a mere 10^{-7} seconds whereupon the universe entered

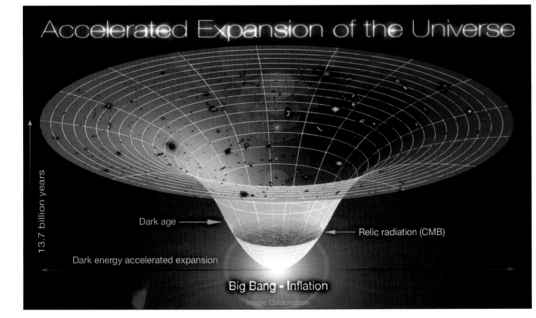

Accelerated Expansion of the Universe

13.7 billion years

Dark age

Relic radiation (CMB)

Dark energy accelerated expansion

Big Bang - Inflation

image: Coldcreation

LEFT Incorporating the physics of dark energy and dark matter, the understanding now is that the Universe is continuing to expand and that the rate of expansion is accelerating, portending a time very far in the future when there will be no more star-making material around. *(ColdCreation)*

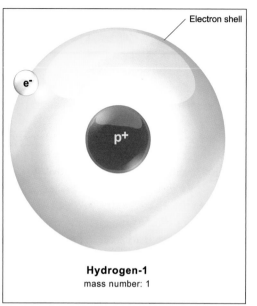

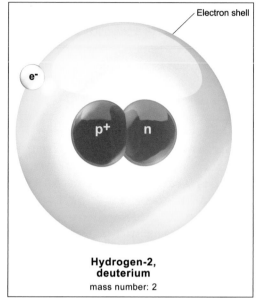

Electron shell

Hydrogen-1
mass number: 1

Electron shell

**Hydrogen-2,
deuterium**
mass number: 2

the electroweak period (10^{-36} seconds after the Big Bang), where the temperature had cooled further to 10^{28}K.

During the electroweak phase the strong nuclear force separated from the still combined electromagnetic and weak nuclear forces and this lasted about 10^{-4} seconds before the start of what cosmologists define as the inflationary epoch, ending after an unknown duration but probably lasting no more than 10^{-1} seconds. The inflationary epoch can be considered as no more than a flash and is the earliest determined point at which the universe as we know it was formed. That is the 'age' of the universe – 13.8 billion years, although the sheer scale of the magnitude in which time was compressed renders that academic.

In this instant, the universe expanded to a linear dimension 10^{26} times its original size, increasing its volume by a factor of more than 10^{78}. The field thus created achieved its lowest energy state, separating out the electromagnetic force as the last of the four known forces – gravity (combining elementary particles according to their mass); strong nuclear (the binding force of nuclei); weak nuclear (radioactive decay); and electromagnetism (a broad spectrum from radio wave to gamma rays and including visible light).

The weakest force is gravity, the only one that combines to multiply its effect, with gravity and electromagnetism declining on the inverse square law. The strong nuclear force operates within the radius of a single proton or neutron.

Within one complete second, from this flash appeared a dense mixture of quarks, anti-quarks and gluons, the building blocks of

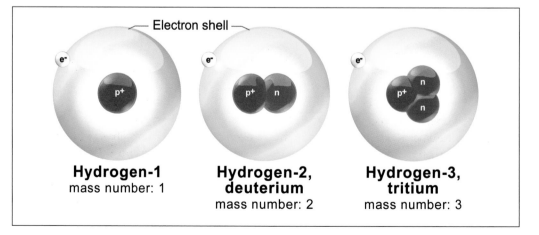

Electron shell

Hydrogen-1
mass number: 1

**Hydrogen-2,
deuterium**
mass number: 2

**Hydrogen-3,
tritium**
mass number: 3

particles from which atoms would form. From this soup emerged hadrons, with baryons typified by protons and neutrons, while neutrinos came apart and began free motion.

At one second after the Big Bang, the lepton epoch began. Within nine seconds of the lepton epoch, the universe experienced the mass annihilation of hadrons and anti-hadrons in a struggle to the death for supremacy, leptons and anti-leptons becoming dominant right across space. The temperature had fallen sufficiently low to stabilise a small residual lepton population. After these 10 seconds the universe entered the photon epoch where light first shone upon the dark ages of primordial events and the first atoms were formed through nucleosynthesis. But it was not light as we know it, the light being re-absorbed by dense plasma as rapidly as it formed.

The photon epoch would last 380,000 years and is itself divided into separate phases where nucleosynthesis, the domination of matter, and recombination takes place in sequence.

Let there be light

Beginning three minutes after the Big Bang and over a period lasting 17 minutes, the nucleic balance of hydrogen and helium isotopes emerged, this consisting of protium (^{1}H), deuterium (^{2}H), helium-3 (^{3}He) and helium-4 (^{4}He). Otherwise known as hydrogen-1, protium is the most abundant hydrogen isotope (>99.98%) and consists of a single proton. Its isotope (deuterium) consists of one proton and one neutron.

In the process of nucleosynthesis, between three minutes and 17 minutes after the Big Bang, atomic nuclei began to form as the temperatures fell, hydrogen ions (protons) combining with neutrons to form the hydrogen isotopes of deuterium through nuclear fusion and the deuterium combining to form helium-4.

As the temperature fell still further, at 17 minutes nuclear fusion could no longer take place, heavier elements beyond helium, lithium and beryllium (such as boron) could not form and almost all the free neutrons had been fused into helium nuclei. Across the universe, by mass there was three times the amount of hydrogen than helium-4 and very few light nuclei remained.

But at this phase there are no heavier elements and there is no 'light' that would be visible to human eyes.

The universe remained in this dark soup of hydrogen, helium and random light nuclei for 70,000 years, when the domination of matter began and the density of the nuclei and relativistic radiation – photons – became equal. It is in this period that cosmologists believe cold dark matter emerged, where gravitational collapse becomes feasible, an essential ingredient in the mix of forces and energies required to make stars and planets.

From this point, dark matter particles interact with each other exclusively through gravitational and weak nuclear forces. Dark matter is one of the least understood, yet one of the most important constituents of matter, accounting for 95.1% of all the matter *known to exist* in the universe, only 4.9% being in atoms and chemicals that make up the stars, galaxies, dust, gas, nebula and plasma we can observe on the electromagnetic spectrum. However, on the basis of the principle of mass-energy equivalence, the dark matter component accounts for 26.8% of the mass-energy structure of the universe while dark energy constitutes some 68.3%. Dark energy is believed to be responsible for the accelerating rate of expansion in the known universe.

Throughout the photon epoch of approximately 380,000 years, the temperature of the universe remained at about 10,000K from where the cosmic microwave background (CMB) radiation comes, a temperature signature that can be measured today. About 377,000 years after the Big Bang, the density of hydrogen and helium atoms started to fall as the temperature of the universe slowly declined.

Protons carry a positive charge and neutrons – by definition – have no charge. At first hydrogen and helium atoms were ionised, because there are no electrons and therefore they carry a positive charge, but as the cooling proceeded electrons were captured to form electrically neutral atoms (the positive proton cancelling out the negative electron).

This recombination epoch is referred to as the decoupling phase, where photons can operate freely without absorption, lighting up the universe for the first time. It is from this period on that the pace of events slowed, but from which more familiar episodes emerge.

Gradually, over the next 150 million to one billion years, a process of reionisation took place as the first stars began to form from gravitational collapse. The intense levels of radiation involved in this process created plasma, which is one of the four fundamental states of matter, others being solid, liquid and gas phases. It can form from heating a gas or by exposing a gas to an intense electromagnetic field, which dramatically changes the quantity of electrons and the formation of positive or negatively charged ions. Plasma is a familiar sight in our world today as lightning and electrical storms, electrical sparks, plasma globes marketed as table attractions or from neon lights – which are in fact plasma lights.

The period after the decoupling phase in the evolving universe is known as the Dark Age, during which the first galaxies began to form around 500 million years after the Big Bang. The simultaneous formation of stars and galaxies powered the engine of the universe within the first billion years and the processes by which the heavier elements were formed required several generations of stars to come and go, increasing the production of materials crucial for planetary systems, including that which contains Earth.

Making stars

What we have been describing so far is *primordial nucleosynthesis*, where the universe begins in a hot and energetic state populated by fast-moving particles and neutrons and where positively charged hydrogen nuclei ($^1p^+$) and neutrons (^1n) exist. But the neutrons are unstable because they have a half-life (defined as the time it takes for half of it to decay) of only 10.3 minutes and only after the temperature cooled sufficiently could deuterium combine with protons, neutrons and other deuterium nuclei to form helium (^3He) and tritium (^3H), which decays into ^3He. This left a

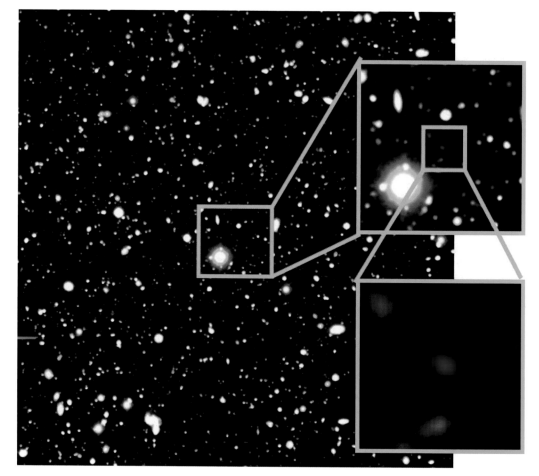

RIGHT Galaxies began to form very early in the Universe, this Hubble Space Telescope image showing the farthest in distance and the furthest back in time, at 12.88 billion light years, observed to date. *(NASA)*

small amount of lithium, beryllium and boron, and within 700,000 years the temperature had dropped to the point where electrons joined the remaining protons and heavier nuclei to form atoms. It is from this matrix of atoms, particles and the four forces that stars began to form.

Scientists call star-forming *stellar nucleosynthesis*, because that is exactly what it is: the conversion of matter into energy through nuclear reactions that transform hydrogen nuclei into alpha particles (two protons and two neutrons bound together in the same manner as a helium nucleus). As the effect of gravity grows through accumulation of matter, increasing quantities of matter are drawn into the centre and this contraction leads to thermal pressure that resists ultimate collapse. The thermal pressure itself results in increased density and higher temperatures in the stellar core. The process continues to grow until no more matter can be drawn in from the local region.

As contraction grows and the temperature rises the core becomes sufficiently hot for thermonuclear fusion to begin, where atoms fuse together releasing great amounts of energy as radiation and light. The rate at which reactions take place increases with temperature. This creates a balance, or equilibrium, between the gravitational force of the assembled stellar mass that tries to contract it to a single point, and the radiation (or thermal) pressure which keeps the mass inflated like a balloon. When the two are in equilibrium the size and brightness of the star is said have entered the stable period of its life, which can last from several million to several billions of years.

Astronomers describe stars from the earliest stellar-forming phase as Population III types: stars that are formed of lighter elements and contain very few metals – because the heavier elements that would be found in later stars had not yet formed. From our perspective, these were the most important events in the history of the universe because they began a train of events that would lead, eventually, to the formation of our Sun and the solar system itself.

Analysis shows that Population III stars which formed at this much earlier, hotter, period in the star-forming age (which persists today) would allow them to be up to 130 times the mass of our Sun. Observations and calculations indicate that several such stars would form together, very large stars hosting several large stars around them. Astronomers hope to be able to observe Population III stars with the James Webb Space Telescope (JWST), designed to fly in 2018 and operate for a while in parallel with the Hubble Space Telescope.

But these stars contained very few of the heavier elements which would form planetary systems. Nevertheless, the material from which stars with a higher metallicity would form was forged in the nuclear furnaces of Population III types when they liberated their products to the mix of gas and dust in space.

Population II stars formed later still, and were similarly depleted in metals, although they contain higher ratios of the alpha elements (oxygen, silicon and neon) compared to their iron content. Stars with even higher metallicity are referred to as Population I types, and include our Sun which, because of the relatively higher abundance of heavier elements, can support the formation of rocky bodies in orbit about them.

Population I types are more usually found in the spiral arms of our galaxy while younger stars than the Sun are closer in to the centre of the galaxy. Population II types can be found in globular clusters and in the halo of stars that surround our galaxy in a spherical canopy. But all population categories have played an important part in building heavier elements with each successive phase and discharging their manufactured products into the regions from which later stars form and from which planetary systems emerge.

Star formation starts when a large mass of gas and dust starts to come together through contraction from a molecular cloud. This can be up to 100 light years across, containing a mass equal to six million times the mass of our Sun 4.39×10^{30}lb (1.99×10^{30}kg). Exactly how the protostar begins depends upon the composition of the molecular cloud and the way in which it will evolve depends upon its mass. This will also determine the conditions under which it evolves.

Normally, interstellar clouds of gas and dust remain in hydrostatic equilibrium while the kinetic energy of the gas is balanced by the potential energy of the gravitational force. If the cloud grows in size and mass the pressure of

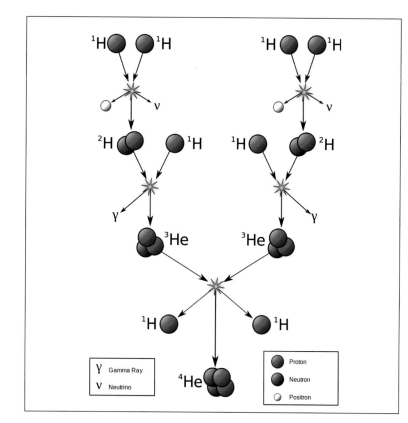

γ	Gamma Ray
ν	Neutrino

●	Proton
●	Neutron
○	Positron

ABOVE The Sun operates on the proton-protonchain, where core temperatures are 106K. Fusion reactions are sensitive to temperature and all stars, of whatever size, operating on the p-p chain vary only between 46K and 406K, despite there being a significantly higher ratio in size between the smallest and coolest and the largest and hottest.
(David Baker)

the gas will be unable to support it and it will begin to collapse, overwhelmed by gravity. The amount of material in such a cloud can be equivalent to several tens of thousand times the mass of our Sun. Many stars can form in such a collapsing cloud, the amount of time taken for them to reach thermonuclear reactions depending on the amount of localised gravitational collapse experienced.

There are other ways of forming stars, including the collision of unstable clouds of matter, exploding stars that create a shock wave to compress clouds of gas and dust, and tidal forces within clusters of other stars. There is no way of knowing which particular trigger caused the formation of our own solar system, starting with the birth of the Sun, but it is strongly believed that our star formed along with several others in the vicinity. They, like us, have taken their own course around the galaxy and now spread far and wide, so there is very little way of knowing precisely which stars were associated with our local group when it formed.

Evidence for the age of our Sun comes from ancient meteorites that show the existence of short-lived isotopes that only form when giant stars explode as supernova. One

of these isotopes is iron-60 and its presence indicates that our Sun formed in the vicinity of a massive explosion that compressed gas and dust to begin the process of collapse. But the birth of our Sun, as well as the processes that produced the materials from which our solar system formed, had their origins billions of years earlier.

The life phases of a star such as our own Sun are typical for all but the biggest and the smallest on the stellar scale of size and longevity, and it all begins with the protostar.

Star shine

The formation of a protostar follows the same physical laws as those associated with the balancing of gravity and the kinetic energy of the gas cloud from which it emerges. As it collapses down the density increases to $19^{-13}g/cm^3$, at which point the core remains stable but the temperature increases. Infalling gases collide with the core and create shock waves that add to the heating phase. At a temperature of about 2,000K, the energy dissociates the hydrogen molecules followed by ionisation of the hydrogen and helium atoms.

As the density of the material falls back below $10^{-8}g/cm^3$ it is sufficiently transparent for radiating energy to escape from the protostar, but this reduction in pressure allows gravity to overwhelm equilibrium and for the collapse to recommence. It continues to do so until a new thermal equilibrium is reached as the hot gas provides the pressure to support the star internally. When the pressures and temperatures are high enough, thermonuclear reactions commence and the star begins to move toward the end of its formative stage.

For a star the size of our Sun, the principal nuclear reaction is on the proton-proton (p-p) chain that begins: $2(^1H + {}^1H \rightarrow {}^2H + e^+ + \nu_e)$, where e^+ represents a positron (the anti-particle to the electron), ν_e an (electron) neutrino. In this, hydrogen fuses to build helium atoms and so on. The reaction rate of the chain becomes great around a nucleic temperature of 3×10^6K and the thermonuclear stability is reached at a core temperature of about 15×10^6K. But even at 15 million degrees, there is still a lot going on to regulate the consistency of the long-

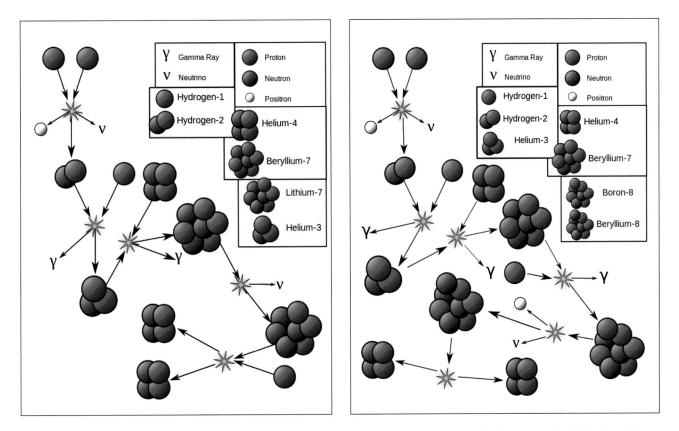

The legend for both diagrams includes: Gamma Ray (γ), Neutrino (ν), Proton, Neutron, Positron, Hydrogen-1, Hydrogen-2, Helium-4, Beryllium-7, Lithium-7, Helium-3. The right diagram also includes Helium-3, Boron-8, and Beryllium-8.

term stability – because over short periods it is anything but stable.

The fusion cycle in the Sun follows four separate p-p chains, one of which has never been observed in the Sun. Before those options are available the first event is the fusion of two hydrogen nuclei into deuterium, which releases a positron and a neutrino with a single proton changing into a neutron. This is a two-step process in which first two protons fuse and form a diproton, producing a beta-plus decay of the diproton to deuterium and the release of 0.42MeV in energy. The half-life of a p-p fusion chain in the Sun is probably around a billion years but there are further reactions.

The single positron emitted by the beta decay is annihilated instantly by an electron and this combined mass energy is removed by two gamma ray photons producing 1.02MeV. The residual deuterium thus formed fuses with another proton to form the light isotope of helium, ^{3}He, a process that happens very quickly and is governed by the strong nuclear force (binding between particles) rather than the weak nuclear force of radioactive decay. In the Sun, the deuterium lasts for a mere four seconds before it is converted

ABOVE LEFT The proton-proton II (p-p II) chain occurs in 14% of reactions within the Sun, where helium-3 and helium-4 fuse to form beryllium-7 and lithium. *(David Baker)*

ABOVE RIGHT Less than 0.01% of nuclear reactions in the Sun follow the p-p III path, where more complex chains link to produce isotopes of beryllium and boron. *(David Baker)*

into He-3. It is from this point that there are four separate paths for the reactions to follow.

In the Sun, 86% of the thermonuclear process follows the p-p 1 path, where two light isotopes of helium fuse to form He-4 releasing 12.86MeV. Only 2% of this energy is lost to the two neutrinos produced. Two other paths are also followed: p-p 2 where ^{3}He and ^{4}He fuse to form beryllium-7 and lithium; and p-p 3 where more complex ^{3}He and ^{4}He fusion chains take place. The p-p 2 path accounts for 14% of reactions in the Sun, with p-p 3 found in only 0.11% of reactions. A possible p-p 4 path exists, where helium-3 reacts with a proton to create helium-4 with the higher energy level of 18MeV, but this accounts for a mere 3 parts per million of solar reactions.

The total energy release of the Sun is 26.73MeV from the 0.7% loss of mass from the original protons during these reactions. Gamma

RIGHT A background shock wave hits a stellar cloud in Orion, typical of the concussive events that surround compressive forces, driving gas and matter into a dense mass, raising gravitational attraction and beginning a starforming process. This is one of several paths from which protostars can begin. *(NASA)*

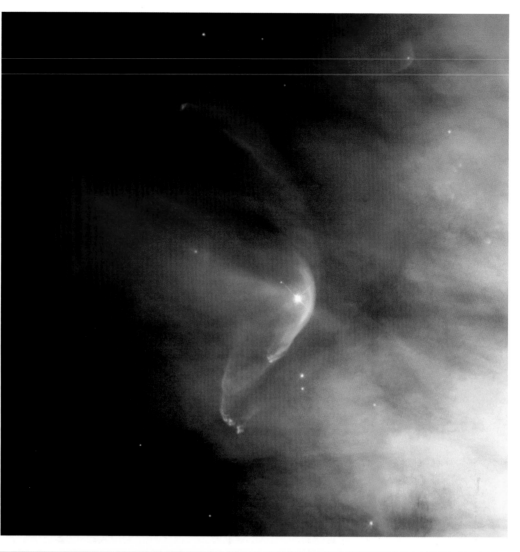

BELOW Situated in our galaxy and a mere 6,500-10,000 light years distant, the Carina Nebula is about 900 light years across, within which are several open star clusters typical of the stars which form in groups, only to drift apart on their separate paths around the galaxy. *(HST)*

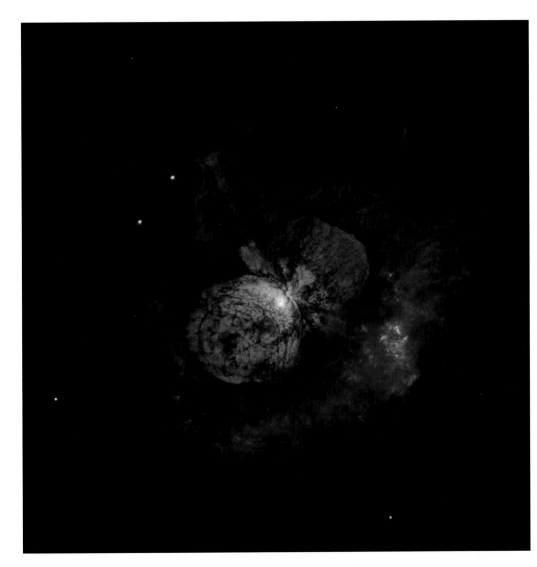

rays released during these events heat the interior of the Sun and continue to prevent it from collapsing under its own gravity.

Stars larger than 1.3 times the mass of the Sun, with internal temperatures of 17×10^9K, convert hydrogen to helium on the carbon-nitrogen cycle. These levels of energy output are typical of the stable life of the Sun after its proto stage but were not as stable during its period of formation.

Energy generation levels fluctuate widely and but for a steep temperature dependence acting as a thermostat, the collapsar, as it is called, would quickly become unstable and never form as a long-lived star. As the core heats up it expands, which causes the temperature to fall, whereupon it shrinks slightly until compression again causes it to heat sufficient for an increase in fusion rates required to hold it up against the inexorable force of gravitational compression.

Fusion rates are heavily dependent on temperature and only a slight increase can cause greatly increased luminosity, but these oscillating periods of core expansion and contraction last a long time.

The stable lifetime of a star depends on its temperature and mass. Astronomers classify stars on this basis using a series of letters further subdivided into numbers. The largest and brightest (O types) are about 120 times the mass of our Sun; 1.4 million times brighter, they have a surface temperature of 53,000K. But they have a very short stable life of around two million years. By contrast, the smallest and coolest (M) types are around 3,000K with a luminosity much less than one-tenth that of our Sun but with a lifetime of 68–700 million years.

These smaller stars are thought to form through gravitational collapse of a rotating agglomeration within molecular clouds. This leads to the formation of a rotating accretion disc – so called because it draws in matter from outside the main zone of collapse – from which additional matter is channelled into the central region of the protostar. Very massive stars have a complex start-up mechanism and this is not clearly understood, not least because there are so few in the universe.

Our Sun is classified as a G2 star, with a surface temperature of 5,800K and a stable life of around 9 billion years. Its birth is typical of stars of its type and began when a stable cloud of gas and dust began to compress into what astronomers call a stellar condensation. These can happen relatively quickly, the collapse taking a mere 10,000–100,000 years from an initial envelope probably 1.5–2 light years across. Added to the complexities of the collapse outlined earlier, the material from which our Sun and its then adjacent protostars formed begins to coalesce around the central region where rotational forces are greatest and the rate of spin is faster than farther out in the cloud.

The shape and form of this cloud condensation is initially like a shrinking sphere, distorted to some degree by the pressures of ultraviolet radiation in the galaxy and by stellar-induced winds caused by energetic events, not least the shells of discarded nebula and the occasional supernova. With the collapse rapidly taking on the form of a disc, material rains down on the central region where stars are forming, and this excites stellar birth. As these stars form, and as our own Sun would have experienced, giant jets of matter are emitted in opposing directions from the rotating collapsars. These bi-polar jets are powerful and help to

RIGHT Devised more than 100 years ago by two astronomers, the Hertzsprung-Russell diagram is a plot of a star's brightness against its temperature, defined by its spectral colour. It is a way of noting how stars form, joining the main sequence as a product of their luminosity and temperature, which also determines their mass. All stars are rated according to the letter sequence O, B, A F, G, K, M (which generations of students have remembered by the lyric "O, Be A Fine Girl Kiss Me!"). Transecting the chart are lines of radius indicating the physical size of respective stars. (NASA)

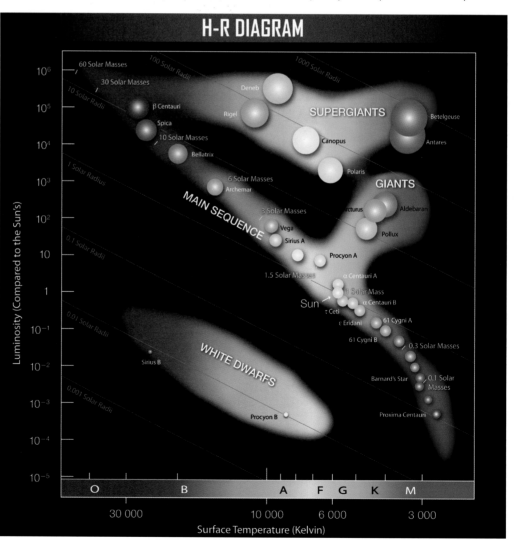

release mass, frequently up to a light year in extent, while the gradually rotating protostar has, as yet, insufficient spin rate to lose material through centrifugal force.

But that spin rate increases according to the law on the conservation of angular momentum – where the amount of energy in motion must be expressed through an increase in the rate of rotation commensurate with a decrease in the radius of the mass. In other words, as the radius of the rotating mass decreases, the spin increases. The centrifugal forces avoided by slow rotation in the initial condensation of the stellar-forming cloud becomes more evident as the star – in this case our Sun – shrinks further. As the spin increases the bi-polar jets clear away the dense fog of gas and dust, at which point the Sun becomes visible.

At this point the Sun is in the T-Tauri stage, where it is a luminous object through gravitational accretion but has not yet begun nuclear reactions. This phase lasts several million years, during which the circumstellar disc appears but with a central cavity, much like the spindle hole in an old vinyl record. This cavity is created as all the separate grains of dust and tiny particles in the disc jostle, collide, lose energy and begin to spiral in toward the emerging Sun. As the particles within the disc get closer to the collapsar (our Sun), they heat up and quickly reach the sublimation

temperature of around 1,500K (1,225°C), and move from the solid to the gas phase.

Conversely, the farther out the grains of dust are the slower they orbit the protostar, occupying a place where the disc is cooler and the motions less chaotic. Beyond a point now defined by the distance four times that of Earth from the Sun, close to the orbit of Jupiter, the temperature would have been so low as to freeze water molecules on dust particles and create an ice line beyond which volatiles would reside. This, as we shall see, was important for the formation of the solar system as it exists today.

Another equally important characteristic of the condensing star is its magnetosphere – a sphere consisting of a magnetic dipole with north and south poles that rotates as a single unit along with the Sun itself. The presence of this intense magnetic field clears the inner part of the disc cavity and causes it to move synchronously with the circumstellar disc itself. These magnetic field lines surround the protostar with a doughnut-shaped structure whereby material streaming away from one pole is looped around and connected back to the Sun at the opposite pole.

As it moves around through the plate-shaped circumsolar disc of dust particles and gas the magnetosphere acts like a conveyor belt and carries up to 90% of the fine grains with it and into the Sun. Amounts vary star by star and we

ABOVE LEFT Clusters of galaxies form to share matter in space and over time. Some drift apart, but dense concentrations of clusters and superclusters are common. (NASA-HST)

ABOVE A supernova in the Large Magellanic Cloud, a small galaxy within the local cluster, dominated by the Milky Way and the much larger Andromeda galaxies. (NASA)

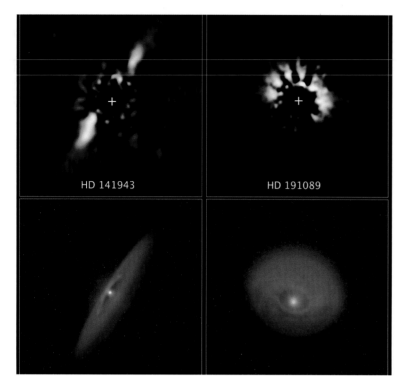

HD 141943 HD 191089

have no way of knowing just what proportion of the dust is transported into the protostar in this way, but it is 70% at the very minimum. But what is happening in this inner portion of the circumsolar disc drives the way the planetary materials will be distributed in the evolving solar system and, as a direct result, what Earth will contain as it forms.

In the inner regions of the disc, intense bombardment from highly energetic particles such as protons and helium nuclei carries

dust grains off along the powerful magnetic eruptions, which accelerate their transport into the Sun. But not all disappear for good, for some of them are blown back out into the disc by the violent force of the jets, or plumes, which recirculate matter into the disc. This intense bombardment of minute grains allows those with neutral electrical charge to fall off into the disc, where they will be deposited at a distance of between the orbits of Mars and Jupiter to a position where the asteroids exist today.

It is from this region that the planets will begin to form, and it is within early meteorites that evidence for this early irradiation is found. For instance, the disproportionately high quantities of ^{26}Mg found in the Allende meteorite is a direct result of it decaying as a daughter product of ^{26}Al, acquired during this early phase. Decaying over millions of years, the refractory inclusions are the telltale signs of collisions and interactions with radioactive elements from which the early evolution of the solar system can be defined. These radioactive nuclei with short half-lives are the result of interaction with the Sun's magnetic field during the T-Tauri stage and validate the accretion-ejection hypothesis.

The powerful eruptions from the proto-Sun, originating in the magnetic field and visible in other stars of this age through their emitted x-rays, are up to 10,000 times as strong as the Sun's magnetic eruptions today and up to 100 times more frequent. But all this violent activity had the result of emptying the circumstellar disc by way of material being swept into

the magnetic fields and to the Sun and by ejection along the rotational disc outside the central cavity well into the outer regions of the disc itself where gravity dominated. The disc, however, is not long lived in cosmological – even geological – time scales.

Studies of other T-Tauri stars going through a similar process to that experienced by our proto-Sun, show the decay of the circumsolar disc taking place over no more than 10 million years, indicating a half-life of 3 million years –

ABOVE This view of the Andromeda galaxy is how the Milky Way would look today, star birth a poor shadow of its former glory as the galaxy winds down and ages. *(NASA)*

RIGHT As our star formed, a dust disc accreted around its equatorial plane, jets of matter shooting out from the solar poles as seen in this star-birth image. *(NASA)*

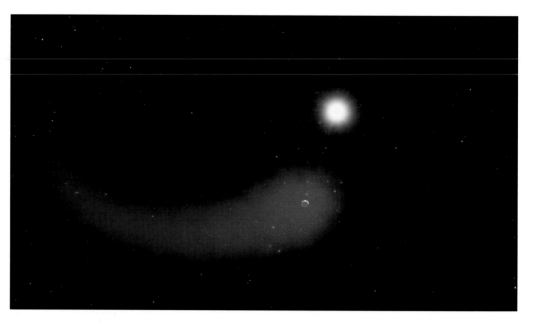

in other words, half of the disc had vanished within 3 million years and half of that by 6 million years, and so on. But there is a dilemma about why the disc should disappear so quickly. The answer lies not in the most likely proposition, that it has all been transported to the proto-Sun or blown out of the solar system, but that it has become something else. And that is the accretion disc from which all the planets and remaining debris left over from the formation of the solar system have formed.

Before we move to that phase, it is important to recognise these precursor stages as providing all the materials and the necessary conditions essential to developing the solar system we live in today. Within 50 million years of the initial condensation phase the Sun had begun nuclear fusion and became an intrinsically luminous star. It had joined the main sequence of its stable life, although the energy output was much less than it is today and the temperature across the solar system was much lower. Now, the events that were to follow would provide the third phase in the evolution of the solar system – the formation of Earth and rest of the planets.

New worlds

Along with this collapsing cloud of gas and dust, and before thermonuclear reactions began, a protoplanetary circumsolar disc evolved in much the same way as the initial cloud condensation, conforming to the same physical laws. In several respects, as we have seen, the laws that governed the collapse of the initial cloud from which all the stars in the Sun's vicinity formed were the same as those that governed the collapse of the local component of that cloud into a protostar and on into the origin of the Sun.

Planetary embryos began to form almost as soon as the collapse of the protostar into our Sun took place, around 4.57 billion years ago. Comprising no more than tiny grains of dust and tiny particles of materials left over from fusion processes in former stars, the circumsolar disc was thin in mass, the Sun itself

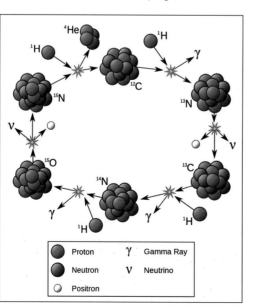

RIGHT Opposing theories as to the precise mechanism by which planets have formed occupied astronomers and planetary physicists for several decades. In the left column is the standard model of accretion and chaotic "settling" after the protostar has become a fully-fledged thermonuclear reactor. In the right column, the gas-collapse model imagines gas and dust coming together through gravitational collapse, inhomogeneous clumps within the gas cloud causing planets to accrete. *(NASA)*

comprising 99.85% of all the mass in the solar system. All the planets, dwarf planets, moons, asteroids, comets, and debris left over from the origin of the solar system, all that would ever form over the next several million years, accounts for a mere 0.15%.

As the accretion process began, the ice regions beyond about 2AU from the Sun began to form the gaseous outer giants, forming first through 'sticky' adhesion at low velocity impacts so as to cause the dust grains and fine particles to glue themselves together through a low-level cold welding. Try placing your hand on a sheet of ice and you will likely lose a thin layer of skin – that's cold welding. At this distance from the Sun icy particles would attach themselves with much greater speed and effectiveness than particles closer to the Sun, where the four terrestrial planets reside today. Dry and hot particles bind together more efficiently through electrical forces of opposing magnetic polarity, and so grow small gravity-wells or agglomerations that increase in mass

RIGHT Multiple star systems forming binary and trinary groups co-orbiting a common centre of mass are common, single star systems such as our solar system being relatively uncommon. Close in to the orbiting pair, gravitational perturbations caused by the stellar masses would destabilise planets and disrupt their orbits. Few planets could be expected to remain stable under such circumstances. Planets much farther away would be more stable and less affected by gravitational inconsistencies but here any such accreted worlds would be cold and receive little light. *(NASA)*

TWO PLANET FORMATION SCENARIOS

Accretion model	Gas-collapse model

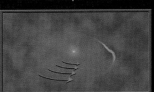
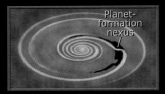

Orbiting dust grains accrete into "planetesimals" through nongravitational forces.

A protoplanetary disk of gas and dust forms around a young star.

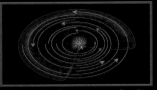

Planetesimals grow, moving in near-coplanar orbits, to form "planetary embryos."

Gravitational disk instabilities form a clump of gas that becomes a self-gravitating planet.

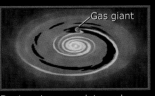

Gas-giant planets accrete gas envelopes before disk gas disappears.

Dust grains coagulate and sediment to the center of the protoplanet, forming a core.

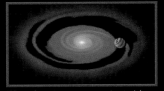

Gas-giant planets scatter or accrete remaining planetesimals and embryos.

The planet sweeps out a wide gap as it continues to feed on gas in the disk.

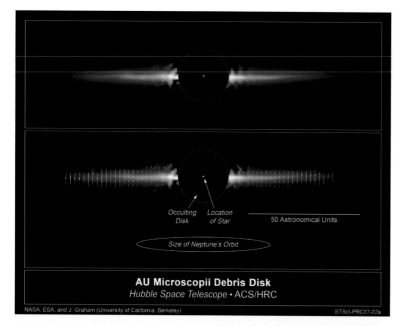

AU Microscopii Debris Disk
Hubble Space Telescope • ACS/HRC

Occulting Disk   Location of Star   50 Astronomical Units

Size of Neptune's Orbit

NASA, ESA, and J. Graham (University of California, Berkeley)   STScI-PRC07-02a

and therefore draw other particles to them, or they attach through collisions and build gravity that way.

It probably took less than a million years for the core of Jupiter to form 20% of its mass, followed by a gradual gathering of light hydrogen and helium elements, pulling more and more matter in upon itself as it grew in size and mass. Jupiter is important because it not only accounts for 71% of all the mass locked up in the planets, but would play an increasingly significant impact on the way the entire solar system evolved, which in turn would affect the way the inner solar system was cleaned out, leaving terrestrial planets like Earth protected from the worst effects of bombardment from debris.

Forming as they did in the ice-zone of the early solar system, all four gaseous giants sucked in a total 99.6% of all the mass in the eight planets, leaving a fractional 0.4% in the four terrestrial planets. And of those four, Earth accounts for more than half the total in the rocky worlds of the inner solar system, with Venus accounting for less than half the total and only a tiny fraction in Mercury and Mars. Put another way, of the four inner planets – Mercury, Venus, Earth and Mars – Earth and Venus have 91.5% of all the mass in those four

LEFT While the distribution of solid material that will form the inner terrestrial planets and the outer giants is consistently the same in random distribution, the less violent areas of the outer reaches allow what will become the gas giants to retain their gaseous envelopes while a debris band forms between the two groups – the protoasteroid belt. *(NASA)*

worlds. For all their importance in the sky and in our observations, Mercury and Mars have very little of the material found in Venus and Earth, both of which dominate the inner solar system, similar to Jupiter and Saturn dominating the realm of the outer giants.

Within about 8 million years Jupiter would have formed to the size and mass it is today – 18 times the mass of Earth. Today it is still contracting under gravitational collapse and shrinking in upon itself, the only world in the solar system putting out more heat than it receives and all of that due to the bleeding-off effect of its initial formation more than 4.5 billion years ago. Saturn would have had a similar formative cycle, as would Uranus and Neptune, although their orbital paths may have been much more erratic than Jupiter and Saturn, which appear to remain stable under all the turbulent conditions at the beginning of the solar system.

It used to be thought that the inner terrestrial planets, forming in the inner, dry part of the solar system, took up to 100 million years to develop to their present size. But that ignores the evidence from the rocks. In setting the date for the birth of the solar system and Earth, scientists are confident that it can be established as 4.568 billion years, the time when refractory inclusions in the calcium-aluminium condensed in carbonaceous chondrites at temperatures above 1,800K.

Earth birth

In attempting to fix a date for the origin of Earth, scientists have used the Acasta gneisses (rocks subject to high temperatures and pressures and usually consisting of bands representing different composition). This rock, which is 4.031 billion years, or just over 500 million years after the origin of the solar system, is the oldest yet found on Earth. But that merely tells the age of the earliest consolidated and surviving rocks discovered to date. Zircon crystals ($ZrSiO_4$) discovered at Jackass Flats in Australia are 4.404 billion years old. While the rocks from which these crystals

BELOW When nuclear reaction begins, a wave of energy blasts through the solar system sending shock waves to clear a lot of the random material and strip protoplanets of their primordial atmospheres. *(NASA)*

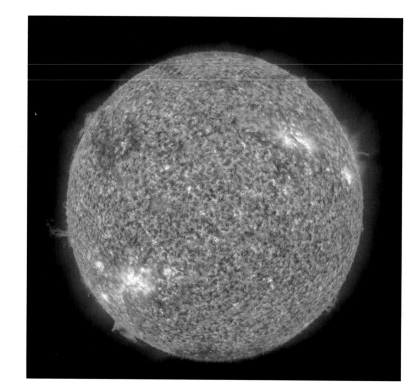

came have long been eroded away, that measurement provides verification of Earth's clock at a mere 164 million years after the planet came through the accretion process, the rapid collection of random debris in the solar system brought together by gravity. Scientists now use 4.568 billion years as the start of the initial epoch, called the Hadean eon, which represents the first 500 million years of Earth, between the zircon crystals and the Acata gneiss.

Geologic time can be confusing – we throw around words without attaching very specific meanings to them. For geologists, the events associated with the changes that have taken place on our planet are divided into separate parallel sections, each a defining segment of time for known changes to the evolution of Earth and its life systems. It starts with the 'supereon', then divides that into the 'eon', followed by the 'era', the 'period', the 'epoch' and the 'age', each of decreasing time span, narrowing the sequence of events.

This geologic timescale also applies to the origin and development of living things, which, particularly after the Precambrian, began to change the face of the planet. It helps refine the sequence while embracing a range of activities related to a common cause. For instance, the four eons have 14 eras which embrace a total of 22 periods. Those periods have 34 epochs, which together have 104 ages, each breaking into greater detail the separate and sometimes not-so-distinct phases. Generally, in this book we will only be concerned with divisions up to and including epochs.

The only application of the supereon is the

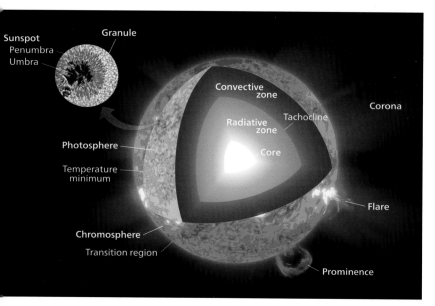

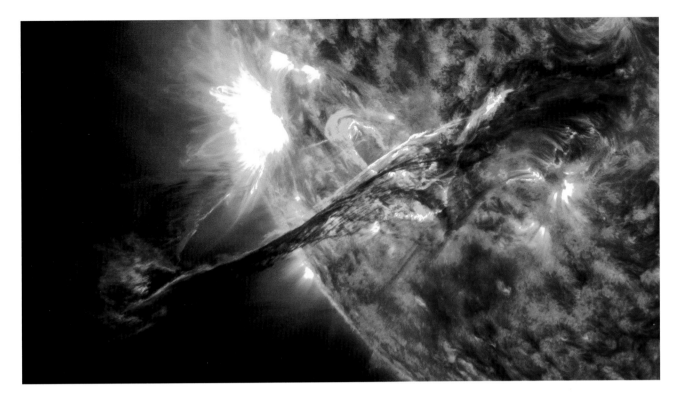

Precambrian, which defines everything from the beginning of Earth to 541 million years ago. The eons, from the origin of Earth to the present, include the Hadean up to 4 billion years ago, the Archean from the Hadean up to 2.5 billion years ago, the Proterozoic from the Archean up to 541 million years ago, and the Phanerozoic from then until the present. As we go through the story of Earth's geologic history, where appropriate we will break those periods down into their separate discrete time segments.

The Hadean eon includes the time during which the planet suffered very heavy bombardment from infalling planetismals, large groups of rock and debris drawn to the planets through gravitational attraction. Planetismals are more like the asteroids we are familiar with today and differed little from them except in size, some measuring more than 500 miles (700km). It is also the time when the Moon formed, and this event had a considerable effect on Earth's early evolution, as we shall see later.

ABOVE A solar prominence extends outward from the photosphere. Cooler and denser than the coronal plasma, prominences can extend out a distance almost equal to the diameter of the Sun and can last for several days or weeks. (NASA)

LEFT A visualisation of Epsilon Eridani with a newly forming solar system containing planets, asteroids and comets, typical of the birth of the solar system and the accretion of the planets. (NASA)

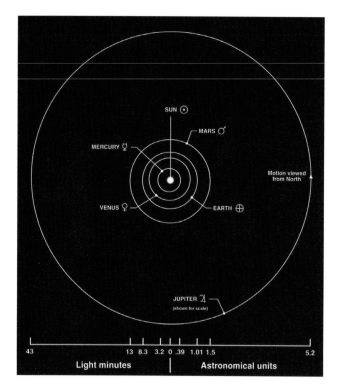

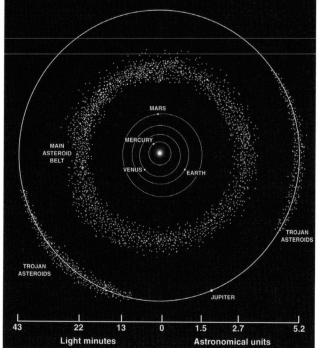

ABOVE As viewed from the north pole of the solar system, the four inner planets and the innermost gaseous giant, Jupiter, for scale, are seen together with the relative distance of each from the Sun in terms of Astronomical Units, being the average distance of Earth from the Sun. The scale of light minutes is the amount of time a radio signal travelling at the speed of light would take to traverse respective distances from the Sun to the planets. *(NASA)*

ABOVE A scale diagram of the inner solar system with the asteroid belt between the orbital radii of Mars and Jupiter showing the positions of the Trojan asteroids in gravitationally stable equilibrium. *(NASA)*

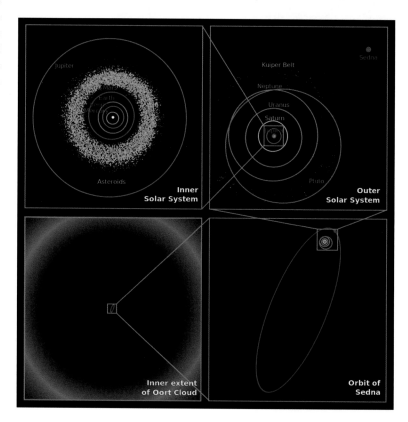

RIGHT A series of reducing scale diagrams show the enormous size of the solar system. Top left: the inner solar system, with the respective positions of the terrestrial planets out to the orbit of Jupiter showing the broad swathe of asteroids; top right: this image is condensed down to the inner component of the previous image to show the eccentric orbit of the dwarf planet Pluto, target for a close fly-by of the New Horizons spacecraft in July 2015; bottom right: this scale is further shrunk to show the highly-eccentric orbit of the Kuiper Belt object Sedna; bottom left: moving still farther away from the Kuiper Belt, this view shows the inner Oort cloud at an even greater distance from the Sun. *(NASA)*

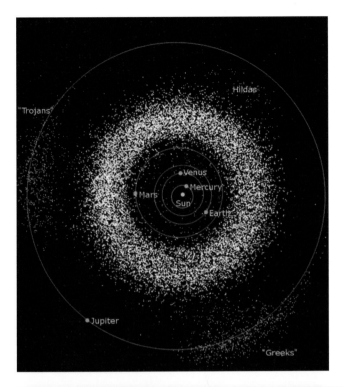

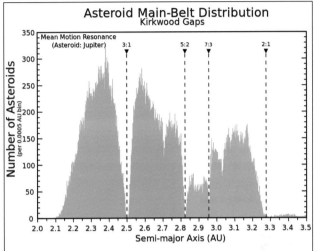

LEFT Separate sets of asteroids are located at the orbital distance of Jupiter and are shown here with the Hilda family of carbonaceous chondrites all of which are locked in a 2:3 resonance with the main planet. *(NASA)*

ABOVE A density/distribution plot shows that the asteroids are located between 2.1 and 3.5AU from the Sun and that they are in resonant orbits with several gaps swept clean by the orbital harmonics of the outer planets. *(NASA)*

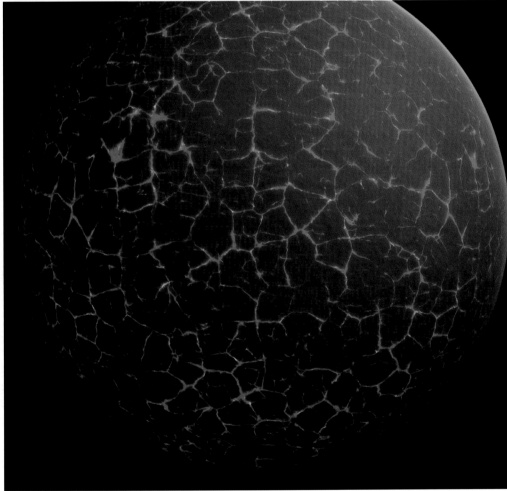

LEFT In the very early stages of Earth's formation its surface was hot, crusted and brittle. Cooling and differentiation subsequently occurred, heavier and denser materials sinking to the centre and lighter materials remaining in the upper layer. *(NASA)*

During the Hadean eon the material forming Earth began to separate out, or differentiate. This process caused the heavier elements to fall toward the centre of the planet and the less dense materials to form a mantle. The core consisted of iron/nickel alloys while the mantle comprised silicates. This differentiation created a heterogeneous world comprising different materials with different densities and mass. As the materials settled out the entire spherical world was molten and this differentiation probably occurred within a geologically brief period of time, perhaps no more than 30 million

ABOVE After cooling quickly, an initial atmosphere of hydrogen, helium and methane would have formed a foggy soup, opaque and boiling at high atmospheric temperature. But it would not last long, as the solar wind stripped it from the planet *(NASA)*

RIGHT The sheer scale of the solar system is difficult to embrace. Here, scale models of the Sun and the planets show the relative size of each object. *(Lsmpascal)*

RIGHT The relative sizes of the four outer giants (Jupiter, Saturn, Uranus and Neptune) compared with the four terrestrial planets (Mercury, Venus, Earth and Mars). *(Lsmpascal)*

years. But what is the evidence for this, and how is it possible to know with such accuracy?

Isotopes of an element are atoms that have the same number of protons and electrons but different quantities of neutrons. When our Sun formed, just like other stars, the process of nucleosynthesis provided different stable and unstable isotopes. Unstable ones decay over time into other elements, and this phenomenon has been used to measure the extinct radioactive element hafnium. Over a half-life of 9 million years, ^{182}Hf decays into ^{182}W, an isotope of tungsten, completely disappearing after 60 million years. It is a particularly good marker because the ^{182}Hf/^{182}W pairs are well adapted to the early phase of the solar system.

Moreover, hafnium and tungsten are good trace markers in themselves because they each have unique geochemical behaviour patterns. Tungsten is a siderophile that has a weak affiliation for oxygen and sulphur

ABOVE RIGHT As the Earth cooled, its surface was bleached by exposure to ultraviolet radiation from the Sun. Quite soon a magnetosphere would form, protecting the surfacefrom the solar wind, but not for several billionyears would an ozone layer block out some of themore harmful UV rays. *(NASA)*

RIGHT Impacts and collisions between bodies orbiting the Sun were common in the very early days of the solar system but all evidence of this very early post-accretion bombardment are erased, only trace chemical analysis indicating that it did occur. *(NASA)*

ABOVE Collisions between planetismals added to the general mayhem in the early solar system as impacts quickly cleared out the inner solar system, the enormous gravitational field of Jupiter helping to attract debris from collisions. *(NASA)*

RIGHT The protoplanet Theia strikes Earth less than 30 million years after it formed, adding 10% to the planet's mass and beginning a sequence that will see the ejected remains form the Moon. *(NASA)*

and tends to bond with iron, so it becomes concentrated within the core. Hafnium is lithophilic, which means that it readily combines with oxygen and forms compounds that do not sink into the core, favouring separation out into the mantle.

If the separation of the accreting Earth into mantle and core occurred after 60 million years all the ^{182}Hf would have been transformed into ^{182}W, and there would be none of this latter element within the mantle. But that is not the case. The excess of ^{182}W found in the mantle could only mean that it had not decayed out by the time Earth had become a differentiated, and cooling body, fixing the separate elements in their dispersed conditions, leaving the hafnium trapped in the mantle to decay into tungsten, where it resides today.

The reason why the metals sank to the core is due entirely to gravity. Being denser, the metals are heavier than the silicates, and most scientists believe that both core and mantle were molten when this took place. Some argue that it could have occurred as a result of metallic liquid percolating between the solid silicate grains, others saying that the sedimentation of the iron into the core took place from two immiscible liquid phases in the magmatic ocean. Whatever the complexity, there is still room for new theories around the basic premise that Earth is in reality a heavily differentiated body.

Today the core of Earth consists of a solid centre and an outer liquid layer, and as the planet slowly cools down from the temperatures it had at accretion and formation as a solid body, the inner core is growing at the expense of the outer core. From calculations based on actual measurement and on theoretical analysis of circumstantial data, it is believed that Earth cooled at about the time the gneiss rocks were found with an age of just over 4 billion years, marking the end of the Hadean eon.

The violence of events during this period is hard to overestimate. The differentiation of core and mantle created great tidal gravitational forces which, had it occurred as a single event, would have raised Earth's temperature by about 1,500K (1,226°C), and this in itself would

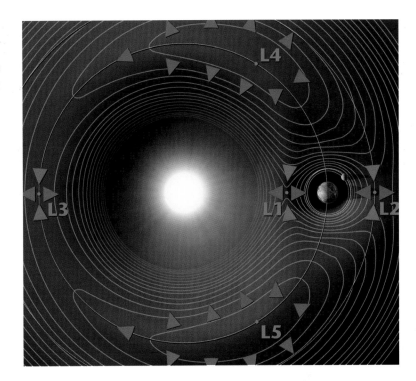

have caused melting of the silicate rocks and the formation of a magma ocean – so the two may be coupled. In fact, examination of rocks returned from the Moon reveals evidence that supports this, and we shall discover more about that later.

Earth has a strong magnetic field, explained in Chapter 3, and a metallic core is essential for it to develop and be retained. It is known by examining marks on the zircon crystals from Jackass Flats that this magnetic field dates back to at least 4.4 billion years, further evidence still for very early differentiation into a rocky core and a silicate mantle.

But even as Earth was coming together and drawing in planetismals from the inner part of the solar system, the planet was on a collision course with another object comparable with that of Mars, half the size of Earth and one-tenth the mass. It would reshape Earth and set the planet on a geologically evolutionary path that may have been the trigger for life, certainly for the fast-track to advanced life long after a crust had formed over the broiling mantle and core. But in studying Earth, why is there such interest in the Moon? Studies of our nearest celestial neighbour, as well as the moons of other planets, add vital detail to missing parts of the puzzle created by trying to unravel the details of Earth formation and evolution.

ABOVE A gravitational contour map of the Lagrangian points L4 and L5 at 60° displacement from the Earth-Sun line. The same gravitational principle that places the Trojan asteroids in positions ahead of and trailing the position of Jupiter in its path around the Sun also determined the orbit of Theia prior to its impact with Earth. *(NASA)*

ABOVE The ejected mass of Theia is hurled back into an orbit of Earth with which it has collided, toppled from its Lagrangian location and soon to form into the Moon. *(NASA)*

Looking to the Moon

The Moon poses several unanswered questions, about its own origin and its evolution. The basic building blocks of the story are well established, not least by the dramatic and sudden surge of scientific data with the Apollo missions of 1968–72, supplemented with results from a range of unmanned spacecraft that have been investigating our nearest celestial neighbour from the early 1960s to the present. But the detail still has many voids, and

scientists are keen to understand the Moon more fully because it provides a window on that early part of the solar system where direct evidence is lacking on Earth.

Due to the churning of the oceanic crust and its continual recycling into the mantle, and because of weathering by a dense atmosphere, physical and material evidence about the early phase of Earth's history are difficult to find on its surface. However, since it is essentially frozen in time the Moon can fill in the gaps, because it covers the first third of

Earth–Moon history and, while the two bodies have evolved along totally different paths, both bodies have been exposed to the same external influences occurring in this part of the solar system since the beginning.

There are many ways in which the Moon directly affects Earth today, and other ways that are inferred from scant evidence. The most obvious is the tidal influence of the Moon on the oceans and seas of Earth. The reason why this is so is discussed in Chapter 4, but the influence on Earth is also felt in the internal structure of the planet, while the gravitational influence of Earth on the Moon has significantly shaped its evolution, at least its internal structure.

The Moon is unique in the solar system in that it is by far the largest natural satellite in relation to its parent planet. With a diameter of 2,160 miles (3,476km) and a mean density of 3.35g/cm³ it is composed of the same density of materials as that of Earth's crust, and the Moon is heavily depleted in water or ice, although there are regions on its surface where ice is thought to exist. Earth, by contrast, has an average diameter of 7,923 miles (12,750km) and a mean density of 5.5g/cm³. Our Moon is not the largest natural satellite, both Jupiter and Saturn possessing some very large moons, each with unique as well as shared characteristics varying widely across the spectrum of size, density and composition. In fact, the four largest moons of Jupiter – Io, Europa, Ganymede and Callisto – were the first objects discovered around that giant planet when Galileo first pointed a telescope at the heavens early in the 17th century.

Only the outer planets have significant moon populations, with a total of 170 already discovered in orbit around the four gas giants, of which 129 orbit Jupiter and Saturn. But of those populations, only four are larger in size than our Moon – three at Jupiter and one at Saturn. The smaller satellites are a mix: large bodies spherical in shape through gravitational attraction, medium-size worlds of complex rounded form, and a large number of asteroid-size objects captured by these giant planets and installed in near-circular or elliptical orbits, some of which are temporary and likely to degrade.

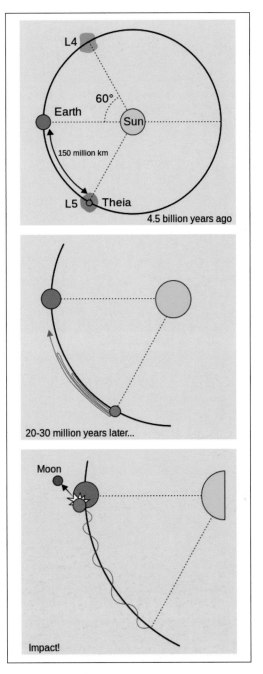

4.5 billion years ago

20-30 million years later...

Moon

Impact!

Studying the moons of other planets helps interpret the early history of Earth's Moon for a better understanding of processes that have reshaped our only natural satellite. This is why scientists place priority on studying the moons of the outer planets and why there has been a great enthusiasm to discover the conditions at and below the surface of these worlds. Understanding the geochemical processes at work inside some of these frigid bodies far out in the depths of the outer solar system advances knowledge about processes

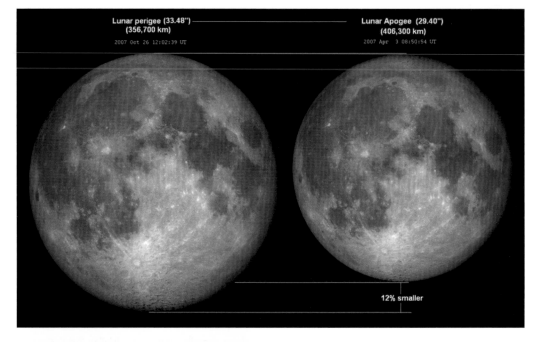

Lunar perigee (33.48")
(356,700 km)
2007 Oct 26 12:02:39 UT

Lunar Apogee (29.40")
(406,300 km)
2007 Apr 3 08:50:54 UT

12% smaller

RIGHT Because the Moon occupies a slightly eccentric orbit it appears to move back and forth, an illusion created by its elliptical path. *(NASA)*

RIGHT The Moon is locked in synchronous rotation with the Earth, one side continuously facing its parent planet. Much of the evidence of early events on Earth has been lost, but the Moon's surface bears testimony to events occurring far earlier than the creation of much of the evidence remaining on the planet. *(NASA)*

RIGHT The far side of the Moon is not as well-known as the near side, only spacecraft making direct visual observations. It bears a very different face to that seen from Earth, covered in ancient craters and with little of the basaltic lava that characterises so much of the near side. *(NASA)*

occurring on our own planet as well as deciphering measurements on both Earth and the Moon.

With a mass 318 times that of Earth and a volume 1,300 times greater, Jupiter hosts four large Galilean satellites (ie those moons originally discovered by Galileo Galilei), which have provided a deluge of information that has completely rewritten the textbooks on planetary moons. Discoveries at Jupiter began with the Pioneer space probes of the early 1970s, followed by the two Voyager spacecraft late in that decade and Galileo, in orbit around Jupiter between 1995 and 2003. Over time the number of known moons of Jupiter has steadily increased as more are discovered, there now being 67 officially known with many more potentially waiting to be unveiled. All but four are small, rocky bodies with insufficient gravity to pull them into a spherical shape, more like asteroids than moons.

Of the four large Galilean satellites, Ganymede and Callisto are larger than Earth's Moon. Ganymede has a diameter of 3,273 miles (5,268km) while Callisto has a diameter of 2,660 miles (4,820km). But Ganymede has a mean density of only 1.94g/cm^3 and that of Callisto is only 1.83g/cm^3. These low density values indicate that there is a higher abundance of lighter elements than in the Moon, with a mixture of rocky material and water ice in equal proportions.

RIGHT In relative size, the Moon is small but in the solar system it is the biggest in proportion to its parent planet, Jupiter and Saturn being akin to mini-solar systems with a prolific abundance of moons, all very much smaller in comparison to their parent bodies. *(NASA)*

Another of Jupiter's moons, Io, is only a little larger than our own, and with a diameter of 2,264 miles (3,644km) and a mean density of 3.5g/cm³ it is seemingly very similar in composition. But in every other respect it is totally different, the tidal influence of giant Jupiter pulling and tugging at its interior to produce volcanic plumes of sulphur and other materials.

Europa has a diameter of 1,940 miles (3,122km) and, with a mean density of just over 3g/cm³, seems not dissimilar in size to our moon, but Europa is believed to have a surface crust covering a deep ocean where primitive life forms may exist in thermal conditions made habitable by internal geological forces and tidal forces from Jupiter.

BELOW A scale comparison shows the Moon eclipsed in size by Io, Ganymede, Callisto and Triton with the majority of moons around the gaseous giants being smaller. Only the major moons are shown here, the planets of the solar system possessing 181 moons of which only 19 are sufficiently massive to have pulled themselves into spherical shape. *(NASA)*

Selected Moons of the Solar System, with Earth for Scale

Earth	Mars	Asteroid Ida	Jupiter	Saturn	Uranus	Neptune	Pluto	Eris

Moon

Phobos
Deimos

Dactyl

Io
Europa
Ganymede
Callisto

Mimas
Enceladus
Tethys
Dione
Rhea
Titan
Hyperion
Iapetus
Phoebe

Puck
Miranda
Ariel
Umbriel
Titania
Oberon

Proteus
Triton
Nereid

Charon

Dysnomia

Scale: 1 pixel = 25 km

Earth

LEFT Earth, Moon and Jupiter's largest moon Ganymede to scale. Some moons in the solar system are like small planets and several are believed to have subsurface oceans, ice and the possibility of life under the surface, where it would be shielded from intense radiation. *(NASA)*

BELOW LEFT A comparative scale drawing showing Callisto, which is significantly larger than the Moon and an excellent choice for a detailed geological survey. In synchronous rotation with Jupiter, it appears to have the oldest and most heavily cratered surface in the solar system – a great deal can be learned from it concerning the early period of the solar system. *(NASA)*

Farther out in the solar system, almost twice as far from the Sun as Jupiter, the ringed planet Saturn boasts another mini-solar system with a gas giant at its centre controlling the orbits of 62 moons. With a diameter of 3,200 miles (5,152km) Saturn's largest moon Titan is also larger than Earth's moon, but with a mean density of only 1.88g/cm³ it is firmly in the realm of the ice worlds such as Ganymede and Callisto at Jupiter. But Titan is interesting in a special way relevant to studies of Earth, since it is the only moon to have a significant atmosphere. Its gaseous nitrogen envelope exerts a surface pressure of 1.45atm (45% greater than that of Earth) with opaque haze layers blocking out all but 1% of the Sun's light.

While most of the atmosphere is nitrogen, direct measurements from Europe's Huygens probe, which landed on Titan in 2004, found 1.4% methane and 0.2% of trace hydrocarbons, probably formed high in the atmosphere by photolysis of methane. Interestingly, all but 5% of Titan's orbital period is within Saturn's magnetosphere, which will protect it from harmful solar radiation. The surface temperature of Titan was discovered to be 94K (-179°C), but conditions on this moon are believed to be close to those on the primordial Earth before life began, about which more in Chapter 8.

Beyond Jupiter and Saturn, of the 27 known moons of Uranus only five are of significant size, all between 294 miles (472km) and

RIGHT Europa is located closer in to Jupiter and the planet's gravitational forces pull and tug at its structure. This diagram shows alternative concepts for what may be a subsurface ocean where, theoretically, life could be sustained.. *(NASA)*

Metallic Core

Cold Brittle Surface Ice

Rocky Interior

H₂O Layer

Warm Convecting Ice

Metallic Core

Ice Covering

Rocky Interior

H₂O Layer

Liquid Ocean Under Ice

LEFT Closest of all the major Galilean satellites of Jupiter, Io is pulled and contorted by the enormous gravitational influence of the parent planet and is as different from our Moon as possible, eruptive sulphur literally squeezing from the interior as its surface moves up and down like that of a trampoline. *(NASA)*

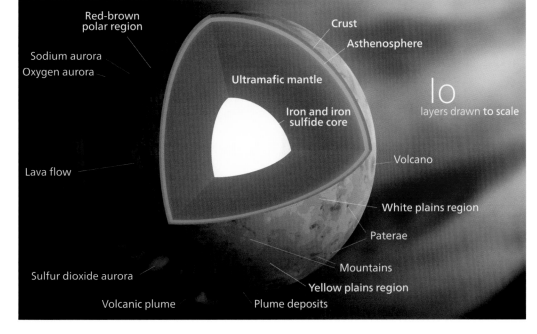

Red-brown polar region
Sodium aurora
Oxygen aurora
Crust
Asthenosphere
Ultramafic mantle
Iron and iron sulfide core
Io
layers drawn to scale
Volcano
Lava flow
White plains region
Paterae
Sulfur dioxide aurora
Mountains
Yellow plains region
Volcanic plume
Plume deposits

LEFT An interpretation of the interior of Io, where extensive surveys by spacecraft enables planetary scientists to build up a picture of a wide variety of conditions on moons throughout the solar system, allowing extrapolation of physical principles at work in extreme environments to inform understanding about the Earth-Moon system. *(NASA)*

ABOVE Saturn's moon Titan has excited scientists for many years, after it was discovered that there may be life on this distant world. A European probe called Huygens was sent to the surface from the NASA spacecraft Galileo, which orbits the planet to this day, sending back images and data. *(NASA)*

979 miles (1,577km) in diameter, none of which are particularly significant for an understanding of our Earth–Moon system. Likewise, the 14 moons of Neptune are relatively small except for Triton, which has a diameter of 1,680 miles (2,705km) and a

RIGHT One of the few images taken on the surface of Titan by the Huygens probe, which survived just long enough to confirm that it has a nitrogen atmosphere with methane-ethane clouds and possibly seas of the same chemistry. It is a place where molecular life could exist. *(ESA)*

particularly interesting history. With a mean density of almost 2.1g/cm³ and an atmosphere of nitrogen, traces of methane and carbon monoxide and a surface more than half covered with frozen nitrogen, it also possesses a strange retrograde orbit. Triton probably originated as a Kuiper Belt object and is only a little larger than Pluto and Eris.

Triton orbits Neptune in a retrograde path, that is, it circles the planet in the opposite direction to the rotation of its parent body, and this is unique in the solar system. Because of this, and the fact that the plane of its orbit lies at an inclination of 129.8° to the ecliptic, it is possible to determine that Neptune captured Triton and that it has not formed out of the condensing nebula from which the planet originated. Triton is locked in synchronous rotation with Neptune, one side always facing the planet it circles, and Pluto is locked in a 2:3 resonance with Neptune, so it is perfectly possible that Triton is a captured Kuiper Belt object.

Through studying the solar system's numerous moons we understand more about the origin and evolution of our own, and by studying the Moon itself we gain data that helps interpret the early phase of Earth's history. But the study of planetary moons can tell us a lot about the development of the inner solar system very early in the history of Earth, and by studying their orbits, and using complex mathematical calculations to retrace their paths early in the history of the solar system it is possible to understand why the Moon is so important to our planet.

While we may be impressed with the size and fascinating geological and atmospheric physics unveiled by several decades of planetary exploration, the outer solar system beyond the asteroid belt remains distinct and separate from the inner region where the four terrestrial planets reside. It is here in the inner solar system that we find a dearth of moons, neither Mercury nor Venus having any natural satellites. Only Mars possesses moons, but these are two potato-shaped orbiting bodies, small and without atmospheres. Many theories have been expounded to explain the origin of theses moons, named Phobos and Deimos, but their precise origin has never been satisfactorily tied down.

From remote sensing and spectroscopy, they appear to have a lot in common with carbonaceous chrondrites, a type of asteroid from the main belt. But the physical means by which Mars could capture these two bodies is difficult to explain. Moreover, with the low mean density of 1.88g/cm³ calculated for Phobos, the structure must be a conglomerate of compacted materials containing voids. Thermal infrared measurement of the surface indicates a preponderance of phyllosilicates, which is more akin to the surface of Mars. It is quite likely that the two moons are the result of an impact with Mars early in its history, and that may carry a fascinating link right back to Earth.

Moon birth

Theories about the origin of Earth's moon go back a long way. Discounting mythical associations and fantastic but implausible ideas saying more about the people who thought them up than anything approaching the possible, the French mathematician Laplace proposed that the solar system began as a giant condensing ring of gas and dust which, while shrinking in size, itself threw off rings, each of which condensed further into what we now know as the planets. He thought that the Moon was a third ejected ring of matter, this one from Earth. We now know that this is physically impossible.

One of the more plausible ideas – the full range would fill a large library – originated with Charles Darwin in the late 19th century. He thought Earth had been subjected to centrifugal forces during its formation when it was molten (itself a radical idea at the time), and that the rate of rotation was so great as to create an elliptical form, bulging at the equator. As this developed, portions of Earth's outer surface were ejected by a combination of gravitational attraction from the Sun and Earth's own period of natural vibration, beginning with a portion of Earth sending out a curving jet of matter which condensed into the Moon and broke free.

This idea was expanded upon in the 20th century by scientists who believed that the place from which the Moon had been torn was the Pacific Ocean. On tearing itself by centrifugal force from one side of the planet, the crust was

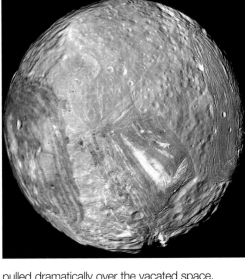

pulled dramatically over the vacated space, tearing apart the crust on the opposite side of the planet, resulting in Africa and Europe being pulled away from the Americas. The idea of moving continents was accurate but totally wrong as to the cause. There is no giant hole in the Pacific, and knowledge about the moving tectonic plates is now well understood as a permanent feature of processes from inside the mantle.

More conservative views had it that the Moon and Earth formed separately in the solar system, very early in its history and at about

LEFT Miranda, a moon of Uranus, has many puzzling features AS one of the smallest to have pulled itself into spherical shape. With a diameter of 292 miles (470km), it has a cracked and scarred surface with giant cliffs 6 miles (10km) high, the largest in the solar system and a reminder of how different the moons and planets are from each other, yet sharing common physical characteristics. *(NASA)*

BELOW At the other end of the scale is the moon Phobos, one of two moons to orbit Mars and characteristic of asteroids and carbonaceous chondrites. *(NSA)*

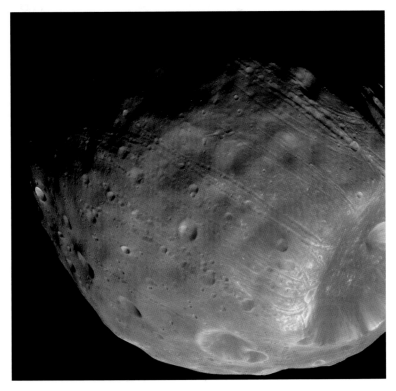

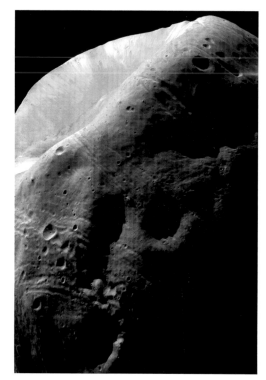

the same time, with the two bodies passing quite close to each other but not colliding. It has been known for a long time that the mean density of the Moon is much less than that of Earth and this was used to indicate that the Moon must have formed from somewhere else in the solar system where there was less iron to condense down into what became the Moon. This low-energy fly-by, it was said, would cause the greater body (Earth) to attract and hold on to the lesser body (the Moon), which put the latter in close orbit around the former, a position from where they have continued to drift apart over the more than 4 billion years that have elapsed to the present.

The argument for the dual accretion case was prescient. It conformed to the theoretical calculations of the French astronomer Édouard Roche, in which there is a minimum distance in which two massive bodies can co-exist before the lesser body breaks up. Had the Moon been torn from Earth via tidal forces, surely, said critics, it could not have passed through the Roche limit zone without itself being torn apart.

The Roche equations hold that the tidal forces of a primary acting upon a secondary body will deform its shape and that the disintegration distance from the primary body will depend, quite dramatically, on whether the secondary is in a rigid (solidified) or fluid state. For instance, in the case of Earth, the Roche limit for a fluid secondary body is a distance of 21,525 miles (34,638km) from Earth, whereas for a solid body it is a mere 5,900 miles (9,492km). In other words, the closest a magmatic Moon consisting of a set of unconsolidated material could remain as a satellite would be 2.88 Earth radii, or 1.49 radii for a rigid Moon. Closer than that and the secondary body would be torn to pieces, forming a ring of debris around the primary.

The effect of the Roche limit can be seen in the extensive ring system around Saturn, a disc of countless tiny rocks, fragments of material from which a moon could have formed, were it not for the presence of the invisible Roche limit. The rings of Saturn have been visible to people on Earth since Galileo pointed a telescope at the sky, but the Voyager spacecraft that flew past Jupiter and then Saturn before visiting Uranus and Neptune discovered ring systems around the other three planets also. It is believed that Saturn's rings, if not those of the other three planets, resulted from the disintegration of a secondary body in close orbit straying into the Roche limit zone. Or, it could have been debris orbiting too close and unable to accrete into a satellite.

But this does not completely discredit the Earth-origin theory of our own Moon, and until lunar samples were returned to Earth beginning in July 1969 many scientists believed this theory may have been correct, although they were unable to explain the method by which this could have been achieved. Direct observation of the Moon has revealed that it is gradually moving away from Earth. Its orbit is getting more distant with time, and that provides direct data for deciphering the origin of the Moon and its future, along with that of Earth. But more of that later in Chapter 8.

It is now widely accepted that the Moon formed as the consequence of a cataclysmic collision between another major proto-planet that had itself formed in the general proximity of Earth. Comparable in size and mass to Mars, with half the size and one-tenth the mass of Earth, this object struck Earth about 4.53 billion years ago, or within 25–30 million years of the accretion of Earth. Scientists have named this object Theia after the mother of Selene, the Greek goddess of the Moon.

It is believed that Theia orbited Earth locked in a Lagrangian point with Earth and the Sun, one of two possible locations which are stable along Earth's orbital path. These positions create a balance between gravitational and centripetal forces. There are several such places, known as Lagrangian points 1–5. Three involve the present Moon, Earth and the Sun, but L4 and L5 are placed 60° ahead of Earth and 60° trailing Earth in its orbital path around the Sun.

Calculations and computer simulated re-runs of the Theia trajectory suggest that when the accreting body of Theia grew to about 10% of Earth's mass this locked position would be perturbed and that any slight deviation in gravitational influence could have caused it to drift away under the influence of Earth's gravity, eventually joining a collision course which may have been caused by its growing mass. There is also an indication that other proto-planets formed and that some collided with each other, Venus being one such example, a world that may have had its own Theia but which, being that much closer to the Sun, would have

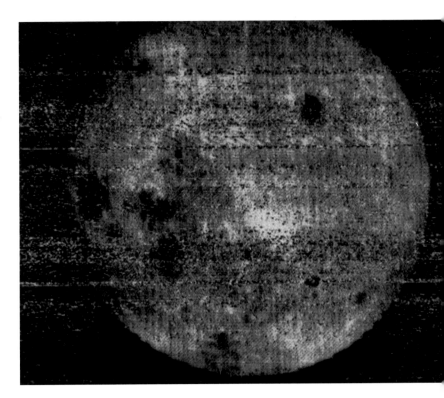

become erratically unstable and may have collided at a more acute angle, burying itself in the outer mantle of that planet and leaving no ejecta field to re-accrete into a moon.

At this very early period in the solar system, the giant outer planets were being reconfigured

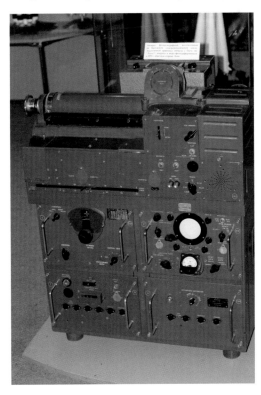

ABOVE The first image taken of the far side of the Moon from Russia's Luna 3 spacecraft in 1959. Although of poor quality it shows how different the two hemispheres of the Moon are, setting aside theories of the Moon from before the Space Age. *(Novosti)*

LEFT The photo-telegraphic system used to pick up the first pictures of the far side of the Moon, sent back as the spacecraft looped back around the Moon for a close pass over Earth. *(Tsiolkovsky State Museum of Cosmonautics)*

to their present positions by gravitational interactions. It is also possible to show that Theia was perturbed in its location with respect to Earth by Venus itself, or by Jupiter, so altering its orbital position that it was deflected to a collision course, striking Earth a glancing blow at an angle of approximately 45°.

The impact would have been less than 8,950mph (14,400km/h), pushing Theia's iron core into Earth's core and spreading its mantle into the impacted planet. Some mantle from both bodies would have been flung out at a similar angle to the approaching trajectory but on the opposite side of the impact zone, much like a long jumper landing on a sand bed would push out material from beyond the front of their foot. This material would have been ejected at such a speed that it would have deposited debris in a cloud around Earth, just outside the Roche limit, between Earth's orbital velocity of about 17,500mph (28,160km/h) and escape velocity, 25,000mph (40,225km/h).

The energy exchange with Earth during the grazing collision would have spun up the planet due to the kinetic moment so that, whatever the preceding period of revolution, a day on the pre-impact Earth would have been no longer than four or five hours. It would also have

placed the disc at an angle to the axis of Earth's rotation. Because of this collision, Earth is tilted in its path around the Sun, but the Moon lies at an inclination of only 5° to the plane in which Earth orbits the Sun – known as the ecliptic. So there is a variation of 18.5° to 28.5° between the tilt angle of Earth in its rotation and the apparent angle of the Moon's orbit around Earth.

Calculations of such an event, together with simulations of the post-impact phase, show that this sequence would have taken less than 27 hours to complete, the impact knocking Earth into its present orbital inclination of 23.5° and placing a disc of coalescing debris just outside the Roche limit. The probability of Theia impacting Earth at sufficient velocity to cause a significant amount if its material to be absorbed into the cooling Earth, while also providing an ejecta blanket of material to be flung out into orbit around Earth but only just outside the critical range of the Roche limit, is very low.

Statistical analysis of solar nebula evolution and the formative stages of planet-building around stars similar to our Sun show that there would be multiple collisions between planetismals and coalescing planets. But the

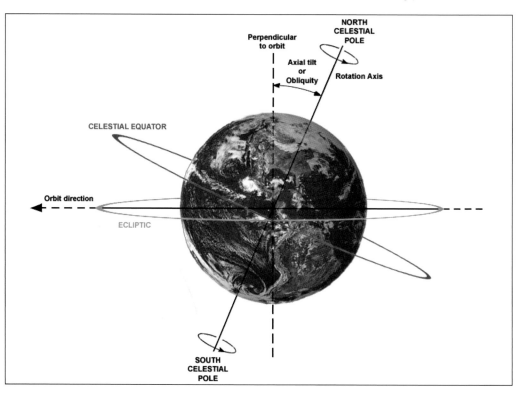

RIGHT Earth's axial tilt, or obliquity, changes over time but averages 23.4° from a line perpendicular to the orbital plane known as the ecliptic, an angle possibly imparted when the Theia planet impacted and knocked Earth off axis. *(NOAA)*

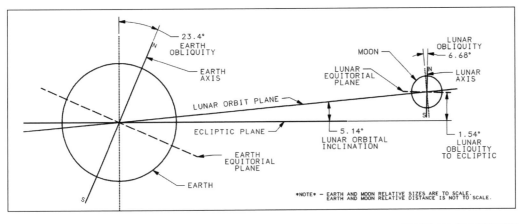

LEFT The plane of the Moon's orbit about Earth is inclined 5.14° to the ecliptic and the Moon's axial tilt is 6.68°, factors imparted during the impact of Theia with Earth and the subsequent accretion into the Moon. *(NASA)*

New Moon	First Quarter	Full Moon	Last Quarter	New Moon

——Waxing Crescent—— ——Waxing Gibbous—— ——Waning Gibbous—— ——Waning Crescent——

precise sequence of events which took place to form Earth–Moon system is highly unlikely. Many subsequent events associated with the development of life and the rate at which advanced life evolved can be attributed to the formation of the Moon and the likelihood of this sequence is difficult to estimate. The fine balance between worlds colliding and destroying each other, worlds colliding and merging, and worlds colliding and producing such a jettisoning of material to a disc orbit from where a secondary body accretes is improbable at best. Yet it is the only theory that fits the evidence.

Moon growth

Evidence from Apollo data, and from calculations based on its mean density, shows that the Moon has about 10% iron, compared to 30% for Earth, which appears to indicate that the collision occurred sufficiently late in its accretion for the impacting Theia to have already differentiated. But the impactor was able to contribute about 10% to the mass of the former Earth, an event which would have thrown quantities of molten rock into the proto-lunar disc at a temperature of about 10,000K (9,700°C).

About 20% of the former Theia was splashed back into space to form the disc, the remainder staying on Earth. Of the total material forming the disc in orbit about 20% was carried up from the terrestrial mantle, about 80% coming from the mantle of the impactor. It is estimated that it took several centuries for the hot disc matter to cool and to begin re-accreting into a proto-lunar sphere, but once it began it would take less than 100 years for the chunks of ejected debris to reform into what we now know as the Moon. None of this transgresses the law of the Roche limit, for the ejected particles would not have been a consolidated body to conform to the equations which require a fluid or solid body to fractionate – that fractionation had already taken place at impact.

ABOVE Because it is in synchronous rotation with Earth (which has 27 'days' to one lunar 'day'), the orbital phases of the Moon determine that the far side is not the dark side, but only hidden to Earth, having an equal amount of light and dark as it makes one full axial rotation in each orbit. *(NASA)*

and its orbiting moon Charon qualify for that recognition. Both are rotating around a common centre of mass. The barycentre of that system, however, is 1,310 miles (2,110km) outside the centre of Pluto, which has a radius of 715 miles (1,150km). This is because of the relative mass of the two bodies, where Charon has almost 12% the mass of Pluto. Our Moon has a mass only 1.2% that of Earth.

While often referred to informally as a bi-planetary system, Earth and the Moon are not, the Pluto–Charon duo being the only example of such in our solar system. But there is more to our Moon that grabs centre-stage in the story of Earth itself. And that concerns events that occurred not long after the formation of the Moon. Before it began to cool the Moon had an unmelted mantle in a solid state with a deep molten magma ocean forming the upper two thirds of the newly formed body. As it began to cool a crust started to form at the surface where the thermal energy could rapidly radiate to space, accumulating plagioclase, a feldspar defined by the quantity of anorthosite present. It is a major constituent of Earth's crust and is the most abundant element in the outer surface material of Mars.

Most characteristic of this early phase in the Moon cooling off was the anorthositic materials which are today key indicators of this early phase in establishing a lunar crust. Below this crustal surface skin the less dense plagioclase would float on top of the mix of pyroxene and olivine, which sank to the bottom of the molten upper two-thirds of the Moon. Anorthosite is found on Earth, originating at two distinct periods quite unrelated to each other, but anorthosite from the Moon is much argued over as it helps identify detail on the very early phase of lunar evolution and provides direct evidence of crustal cooling phases now lost on Earth due to its more dynamic geologic history.

There is an important analogy here with Earth and the way its crust evolved. Because the Moon is only 1/80th the mass of Earth, the surface gravity is only 1.62m/sec^2 (5.3ft/sec^2) versus 9.78 m/sec^2 (32.09ft/sec^2) on Earth. Because of this the pressure inside the Moon increases with depth more slowly than on Earth, where plagioclase is stable to depths of only 19 miles (30km). On the Moon it is stable down to a depth of about 112 miles (180km). Crystallisation occurrences of the Moon's early magma ocean would have generated large quantities of plagioclase that accumulated through flotation, producing the thick anorthositic crust. This produced an insulating layer through which heat could only pass by way of conduction, which, because it is less efficient than radiation or convection, would have slowed the rate of cooling.

The primary constituents of the lunar crust, exposed today as the lunar highlands, while being largely anorthositic in composition, contain oxygen, magnesium, silicon, iron, calcium and aluminium elements. Traces of titanium, thorium, potassium and hydrogen are present in a crust that varies in thickness between the near side and the far side but has an average depth of 30 miles (50km) – much thicker than Earth's crust. From studying lava released from the mantle later in its evolution, the mantle is known to contain basalts of the minerals olivine, orthopyroxene and clinopyroxene with a much greater percentage of iron than Earth. The mantle consists of three layers, upper, middle and lower, representing the bulk of the Moon from a

BELOW Dave Scott parks the Lunar Roving Vehicle near Hadley Rille on the edge of the Mare Imbrium, a collapsed lava tube where subsurface streams of molten basalt ran just below the surface until cooling and shrinking caused it to collapse. The six expeditions to the surface of the Moon contributed a wealth of valuable data from which the history of the Earth has greatly benefited. *(NASA)*

distance of 215 miles (350km) from the centre out to the crust.

The core is believed to be small, with a diameter of no more than 435 miles (700km), of which the solid inner core has a diameter no greater than about 200 miles (320km), of which about 20% will have crystallised. The precise configuration of the fluid outer and solid inner cores is contested but this generalised description fits most evidence. Under this assumption, the liquid outer core probably has a density of $5g/cm^3$ with probably 6% sulphur in weight with a temperature of about 1,650K (1,377°C).

Dating the Moon

The Moon is covered with the evidence of a phase in the early solar system when planetismals and asteroid bodies collided with the planets and scarred their surfaces. Some impacts came close to shattering these primitive worlds and their evidence remains in the form of basins – giant craters so large that they gouged out significant proportions of the crust, sometimes penetrating down into the mantle. Most are observed on the near (Earth-facing) side of the Moon, but there is one that defies explanation. It lies just around the observable disc of the Moon, but it is a baffling puzzle as to just what caused it, when, and why it is chemically different from either the crust or the mantle, presumably into which the impactor penetrated.

The Aitken basin is the oldest, and the largest on the Moon, covering an area 1,600 miles (2,500km) in diameter and with a depth of almost 43,000ft (13,000m). Because it is on the far side of the Moon it cannot be seen from Earth, except for a small portion of an arcuate rim that rises above the limb of the Moon (the edge of the disc as we view it from Earth). This feature has been named the Leibnitz Mountains and is a regular viewing feature for amateur astronomers and professionals alike. As the reflected light from the Sun reveals this and the shadows lengthen at the terminator (the twilight zone between light and dark as the Moon slowly rotates in its orbit around Earth), the undulations taunted astronomers until detailed photographic evidence from the far side revealed Aitken in all its glory.

LEFT An anorthosite sample composed of anorthite brought back from the Apollo 15 mission and hailed as the "Genesis Rock" due to its age at just over 4 billion years, far older than any complete rocks on Earth. (NASA)

The second largest basin identified on the Moon is situated on the near side, covered now by basaltic lava flows that filled in the terraced slopes after it was formed. Named Mare

BELOW A lunar olivine basalt from the Apollo 15 Hadley-Apennine site, selected for its location close to one of the most dramatic impact events on the Moon. The sample crystallised 3.3 billion years ago and is rich in pyroxene, with 31% plagioclase and 13% olivine. It had been on the surface for 25-35 million years and was collected by astronaut Dave Scott in August 1971. (NASA)

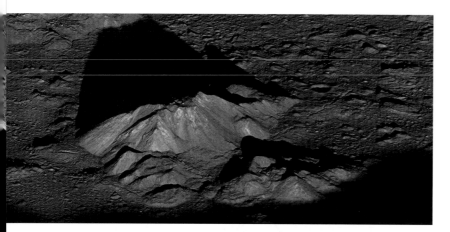

ABOVE **The central peak in Tycho, revealing the rebound feature typical of large impact craters.** *(NASA)*

BELOW **A close-up of the central peak area of Tycho, with a 300ft (100m) diameter boulder in clear sight.** *NASA)*

conjecture in recent years as to the applicability of Copernicus to be typical of bright ray craters, but in general the classification alludes to a much quieter period in the Moon's history than any which went before.

Many events associated with the dramatic origin and early history of the Moon are relevant to a study of Earth's history, and to the practical understanding of why it is unique today, and none could be as game-changing in our ideas about the early Earth than the great cataclysm in the period up to the end of the early Imbrian.

More usually referred to as the Late Heavy Bombardment, this period saw a rain of asteroids, comets, rocks, boulders and debris on a scale hard to imagine. This material was residual left-overs from the general accretion of the primary planets and from impacts and collisions that blew apart early worlds forming secondary bodies themselves shattered asunder by subsequent impacts.

The LHB appears to have occurred across the inner solar system, affecting Venus and Mercury as well as Earth and Mars. But studies of the Moon, and findings from lunar samples, substantiate this bombardment on other worlds as well as our own. This has dramatic implications for interpreting the early period of Earth's history, and data from spacecraft at Mercury appears to corroborate the view that it affected all the terrestrial planets. Radiometric dating of lunar samples shows that melt events occurred at all the sites visited and that it was on a truly global scale. This information will help us interpret the events on the early Earth and help explain why it is so different today. It has been a revolution in Earth science as well as our understanding of the Moon.

For thousands of years knowledge about the Moon was scant and based on eyeball observations, and only in the last four centuries on measurements from Earth using telescopes. Until the space age brought tools, and people, to the surface of our celestial neighbour, and before this new age of exploration, scientists were divided as to the cause of the craters seen across the Moon's entire surface. From early times the dark areas were given names associated with water, since it was, perhaps logically, believed that the moon held great seas and oceans. Only when telescopes began to show large numbers of craters across these areas too did scientists recognise the unique nature of the surfaces to which they had attributed a more Earth-like context.

Up to the 1960s opinion as to the origin of the craters was divided between those who believed them to have been formed by impacting asteroids and debris left over from the formation of the solar system and those who thought they were volcanic in origin. Amateur and professional astronomers alike engaged in fierce debate, polarised around these two

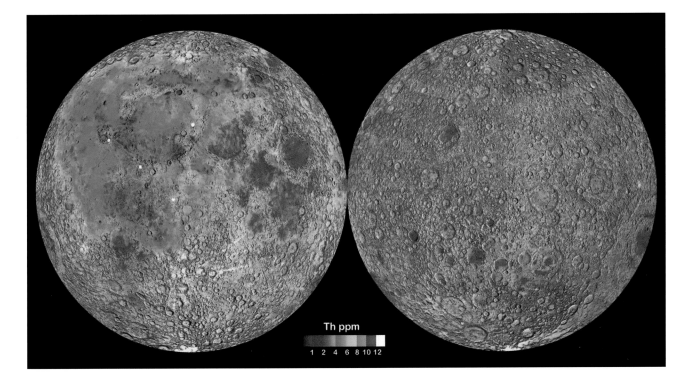

Th ppm

1 2 4 6 8 10 12

interpretations. By the end of the 1960s it was patently clear that craters were of impact origin and that in them lay a code to the age of the event displayed as a scar across the surface. One secret seemed to be in the rays emanating from some bright craters in particular.

Crater rays are formed by splatter material which forms from the lighter fragments thrown out from the surface when the crater is formed by impact. This ray material can extend right around the Moon itself, if formed from a particularly big impact. There are approximately 300,000 craters with diameters exceeding 0.6 mile (1km) on the near side alone and a far greater number on the other side of the Moon hidden from Earth. But the very nature of the way craters form is a mechanism for exploring the segments of Earth–Moon history before the oldest rocks on our own planet.

Ejecta blankets are formed from material excavated close to the surface and to a depth depending on the distance from the rim. The rim itself is from the deepest part of the crater, material that tends to fold upward and outward, while any central peak that forms is a result of the rebound from the elastic crust where the shock waves are reflected back from the circumferential rim and focus back to the central point. Very large craters and basins will

not necessarily have a central peak, the shock waves in these cases being sent scattering across at diverse angles deflected by differing densities in the microseconds after impact.

Samples retrieved from the rim of a crater are from the greatest depth excavated by the impact and therefore perhaps the oldest from that area. Material from the ejecta blanket overlies older material on to which it has been deposited (superposed), while material from far away and on the bright rays comes from near the surface adjacent to the point of impact. Because of this, explorers on the Apollo missions could aim to retrieve samples from the rim of a large crater, move some distance away and take a core sample through the thin layer of ejecta material and down to the original material on to which the crater debris had been superposed.

This is classic stratigraphy, adapted to sample retrieval for age-dating – obtaining a date for the impact event and of the underlying material on which that took place. Not only that, the petrology (origin, composition, mineralogy and history) of each sample can provide a fixed diary of lunar events for that area. The bulk chemistry of the Moon, when traded against local events and precise understanding of the petrology, helps construct a consolidated

ABOVE Geochemical mapping of the Moon reveals evidence of thorium concentrations indicative of basaltic lava plains with quantities in parts-per-million showing that the far side has very few lava floods. *(NASA)*

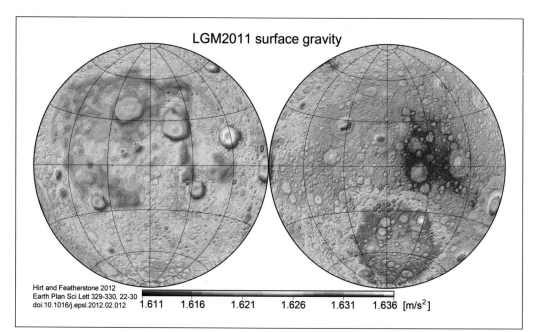

LGM2011 surface gravity

Hirt and Featherstone 2012
Earth Plan Sci Lett 329-330, 22-30
doi:10.1016/j.epsl.2012.02.012   1.611   1.616   1.621   1.626   1.631   1.636 [m/s^2]

picture of the Earth–Moon system before evidence was left on Earth of episodes we would not have known about. That has directly controlled understanding about the first 500 million years of Earth's history.

The production of a thick anorthositic crust was made possible by the lower gravity of the Moon inducing less pressure at greater depth than on Earth. This insulated the thermal flow from the deep interior of the Moon and deferred the period of lava flow into and across the basins over a more protracted period than would have been possible for a body the size of Earth. The basalts of the lunar maria were produced not directly as a result of this delayed thermal flow but to the production of heat from radiogenic elements such as uranium, potassium and thorium. These radioactive materials produced heat from their isotopic decay which was added to the energy imparted from subsequent impacts, albeit less than that during the Late Heavy Bombardment.

Oceanus Procellarum supports the largest volcanic lava flood plain on the Moon. The underlying floor of the flood plain may have been the product of a series of basins that are deeply eroded today and yet still bear signs of their rim structures. It is estimated from seismic profiling and from measurements of the gravity field in the vicinity that the flood plain is in places up to 15 miles (25km) deep. In extent it occupies an area approximately 2,000 miles

(3,200km) across, and some scientists had speculated that it was the remains of a single massive impact that created a basin in which the rim's structure has been largely eroded by the lava flow.

Believing (correctly) that what we learn about the Moon helps us understand Earth better, ever since spacecraft began orbiting the Moon in the 1960s scientists have been intrigued by spacecraft apparently speeding up and slowing down over certain areas of the Moon. While it might appear logical that gravity has an equal force defined by the mass of the body in question, in reality the gravity field varies according to the composition of the planet and of the density and the mass of material directly below. This is so on Earth as well.

Because planets and moons are made from lumps of material accreted from a range of elements, they are themselves lumpy bodies and not homogenous worlds perfectly spherical and of regular consistency. Because gravity is a product of mass, as the mass varies across the surface of the Moon so will the gravity field vary as measured directly on a line from the orbiting spacecraft. The accelerations produced by such gravity variations will cause the spacecraft to speed up or slow down, affecting the overall geometry of its orbit. These regions are known as mascons (maximum concentrations of matter) and minicons (minimum concentrations).

They became an important factor that had

to be incorporated into Apollo flight trajectories, where variations in speed and deviations in predicted orbits could turn possible success into certain failure – not necessarily causing a spacecraft to crash into the Moon (the variations are far too subtle for that) but resulting in the misalignment of flight paths during rendezvous when coming back up from the surface. Unless compensated for, over time these minor perturbations can cause the flight path to vary dangerously in altitude away from predicted values. On at least one occasion sleeping astronauts in Moon orbit had to be awakened to perform a manoeuvre to raise the orbit, as it was being nudged lower and lower on each successive pass.

A NASA mission to study the gravitational perturbations caused by differences in density within the Moon has mapped subsurface troughs in the Oceanus Procellarum which appear to be the result of shrinking of the crust, below the surface, as the magma cooled. The origin of such a system, forming in fact a giant rectangle that encompasses all the dark connected lava plains across the face of the Moon, appears to be defined from an outline traced in gravitational peaks caused by a massive lava plume within the mantle. This has outstanding significance for understanding the interior of Earth where, as we shall see later, there is evidence for multiple plumes on our own planet.

We saw earlier that the averaged-out gravity field of the Moon is 1.62m/sec^2 (5.3ft/sec^2) and in 2011 the two NASA GRAIL (Gravity Recovery and Interior Laboratory) spacecraft were able to sit in orbit around the Moon and observe these anomalies, particularly those of the far side, where the gravity field had been poorly mapped. The measurements show that the variation amounts to 0.0253m/sec^2, which is about 1.6% of the overall gravitational field of the Moon – a very significant amount and one which, when mapped, allows scientists to 'see' what is beneath the surface in bulk and mass, since mass and gravity are irrevocably connected.

Summary

- Understanding Earth requires many science specialists in a broad range of disciplines.
- In the beginning, time and space were formed at the same point from which the universe emerged.
- All matter from which the solar system was made was produced over a preceding period of 9 billion years.
- Star formation occurs in clusters, producing individual suns that drift apart in their galactic orbits.
- Proto-stars draw in material from interstellar clouds that accrete to build planets.
- All the outer planets have moons but Earth's Moon is disproportionately large in comparison.
- The Moon formed out of a collision between Earth and a giant proto-planet called Theia.
- Moon exploration has opened a previously closed window on Earth's early history.

BELOW A poorly defined area known as Oceanus Proecllarum contains many Imbrium-age basins and has been studied to determine the rectangular shape of its structure. It points to a complex origin and one in which a sequence of impacts immediately preceded the Imbrian/Nectarian events. (NASA)

Chapter Three

What drives Earth's engine

Building blocks of rocks

To find out what Earth is made of and why it works as a dynamic and life-supporting planet it is helpful to go back and relook at the structure of matter, and to explore the reasons why different materials exist and what part they play in the evolution of our planet. We examined some of this when considering the origin and evolution of nuclear fusion reaction in the protostar and our evolved Sun.

OPPOSITE Calcium is an alkaline metal, the fifth most abundant on Earth's crust. As a metal it is reactive and harder than lead. Here, travertine terraces at Mammoth Hot Springs, Yellowstone National Park, show off the rock's white, chalk-like appearance. Travertine is a sedimentary rock formed from calcium carbonate minerals. *(David Monniaux)*

Group→1	2	3	4	5	6	7	8	9	10	11	12	13	14	15	16	17	18
↓Period																	
1 H																	2 He
3 Li	4 Be											5 B	6 C	7 N	8 O	9 F	10 Ne
11 Na	12 Mg											13 Al	14 Si	15 P	16 S	17 Cl	18 Ar
19 K	20 Ca	21 Sc	22 Ti	23 V	24 Cr	25 Mn	26 Fe	27 Co	28 Ni	29 Cu	30 Zn	31 Ga	32 Ge	33 As	34 Se	35 Br	36 Kr
37 Rb	38 Sr	39 Y	40 Zr	41 Nb	42 Mo	43 Tc	44 Ru	45 Rh	46 Pd	47 Ag	48 Cd	49 In	50 Sn	51 Sb	52 Te	53 I	54 Xe
55 Cs	56 Ba	*	72 Hf	73 Ta	74 W	75 Re	76 Os	77 Ir	78 Pt	79 Au	80 Hg	81 Tl	82 Pb	83 Bi	84 Po	85 At	86 Rn
87 Fr	88 Ra	**	104 Rf	105 Db	106 Sg	107 Bh	108 Hs	109 Mt	110 Ds	111 Rg	112 Cn	113 Uut	114 Fl	115 Uup	116 Lv	117 Uus	118 Uuo

*	57 La	58 Ce	59 Pr	60 Nd	61 Pm	62 Sm	63 Eu	64 Gd	65 Tb	66 Dy	67 Ho	68 Er	69 Tm	70 Yb	71 Lu
**	89 Ac	90 Th	91 Pa	92 U	93 Np	94 Pu	95 Am	96 Cm	97 Bk	98 Cf	99 Es	100 Fm	101 Md	102 No	103 Lr

ABOVE The periodic table lists all the known elements by group and period displaying the order by atomic number, which determines the number of protons in the nucleus, arranged in rows so that elements with a similar chemistry occur in the same vertical column. As a rule, elements of each category have a similar chemical valency, which is a measure of the number of bonds each element can form. The elements on the table are shown with their chemical symbol, above which is their atomic number. *(USGS)*

tTo recap, the atom consists of a nucleus containing protons and neutrons (except for the hydrogen atom, which has only one positively-charged proton). The proton has mass and a negatively charged electron. For all reasonable considerations the electron has no mass. The proton is in the nucleus and the electron is in a virtual cloud at some undetermined position around the nucleus. The number of protons in the nucleus determines the atomic number and the chemical element, but the negatively charged neutron, when added to the nucleus, determines the isotope of that element.

The atomic mass of an element is the sum of the masses of the proton and the neutrons, the electrons – because they have no mass – being irrelevant here. Carbon-12 has six protons (and six electrons) and six neutrons, but isotopes of carbon-12 (an atomic mass of 12) can have seven or eight neutrons, giving atomic masses of 13 and 14. These isotopes of carbon, like all the elements, are never whole numbers. The atomic mass of carbon is 12.011 because the isotope carbon-12 is the overwhelmingly abundant one, but variable isotopes collectively provide the total atomic mass. They each

behave differently. Carbon-12 favours photosynthesis in which carbon compounds are produced from inorganic carbon compounds – important for carbon-based life.

In other examples, hydrogen has an atomic number of 1 because there is one proton in the nucleus, but it has an atomic weight of 1.00794 because there are isotopes of hydrogen (deuterium and tritium) that prejudice the whole number value. Helium has an atomic number 2 (one proton and one neutron) but an atomic weight of 4.002602, again because of the isotopes, helium-3 and helium-4. These isotopes were important in building the elements from which star formation could begin. The next element, lithium, has an atomic number 3 but an atomic weight of 6.941. Hydrogen, helium and lithium were the three elements that materialised at the Big Bang (see Chapter 2.

With an atomic number of 4, beryllium is the lightest element to have been formed as a result of supernova, massive stars, short-lived and ending their days in a massive explosion shedding material from which other stars are formed and from which the building blocks of worlds will be formed. It has an atomic mass of 9.012182, with five neutrons to four protons (^{9}Be) in its stable form. There is a trace isotope ^{7}Be with a half-life of 53 days, and trace isotope ^{10}Be with a half-life of almost 1.4 billion years. Further up the scale, iron has an atomic number of 26 ($_{26}$Fe) and an atomic mass of 55.845 and seven stable isotopes with between 28 and 34 neutrons.

At the other end of the scale, uranium has an atomic number of 92 ($_{92}$U) and an atomic mass of 238.02891, of which ^{238}U is the most abundant, with 99.274% being of this isotope and 0.72% being ^{235}U, the nucleus having 146 neutrons in addition to the 92 protons. Uranium is perhaps the most dramatic example of the energy locked up within Earth's rocks. It has been fundamental to the geologic processes underpinning the planet's history and it has been used to produce energy and liberate its potent power for peaceful and warlike purposes alike. It has been used to power the electricity grid and it has been used to make an estimated 80,000 nuclear weapons manufactured since the mid-1940s.

Uranium is a silvery-white metal in the actinide series, a group of radioactive elements of which actinium ($_{89}$Ac) is the first in the series, followed among others by thorium ($_{90}$Th), which is another important actinide in the development of the early Earth. About 15lb (7kg) of U-235 can be used to make an atomic bomb. It can also be used to make plutonium ($_{94}$Pu), a process that begins by the ^{235}U acquiring a neutron. Uranium-235 is the only naturally occurring fissile isotope and because it is fertile it can be transmuted to plutonium-239 in a nuclear reactor. Uranium-238 has a low propensity for spontaneous fission but it can be induced to react with fast neutrons, while uranium-233 and -235 has a higher fission probability with slow neutrons. Both play a role in sustained nuclear reactions in power stations.

Within the last several decades it has become known that uranium-235 can become active spontaneously, causing natural nuclear reactions. Such a frightening occurrence appears to have taken place about 1.78 billion years ago when 3% of the total uranium on Earth was of this isotope. Sustained nuclear chain reactions are believed to have taken place underground in Gabon, West Africa,

when deposits rich in uranium were saturated with groundwater that acted as a neutron moderator, inducing a chain reaction. The heat generated caused the water to boil, which acted to slow or quench the nuclear reaction.

These reactions have been identified at 16 separate sites in the region, all lasting over several hundred thousand years, producing up to 100kW of energy during this period. A large factor in these naturally occurring nuclear reactions on Earth was the increasing amount of oxygen in the atmosphere, which would have allowed the uranium to be dissolved and transported with the groundwater where high concentrations could build up. Fortunately for us today, these reactions could not now take place. Uranium-235 has a half-life of 700.04 million years, much shorter than uranium-238, and has decayed out from the 3% then to around 0.7% now so that reactions are not possible without the use of heavy water, water which contains larger amounts of the hydrogen isotope deuterium, or graphite.

The fundamental building block, the process by which atomic nuclei grow, is the essential base upon which are assembled all of the elements. The structure of the atom determines the chemical reactions that are possible. The atomic number (the number of protons in the nucleus) provides a calibrated sequence of increasingly heavy elements and these are laid out through a periodic table that organises them according to the atomic number.

It is the distribution of elements that determines the physical form and structure of the universe and governs the nature of stars but it is in the minerals that the structure of Earth is defined. Minerals are the naturally occurring substances from which rocks are formed, and they are defined by their chemical formula and atomic structure. To date, there are more than 4,660 known minerals, and their diversity and relative abundance are controlled by the chemistry of Earth. Minerals are categorised by chemical composition and crystal structure, properties governed by the assembly of chemical components and species of type. They are shaped by the geological environment in which they form and are influenced in that by temperature, pressure and bulk composition.

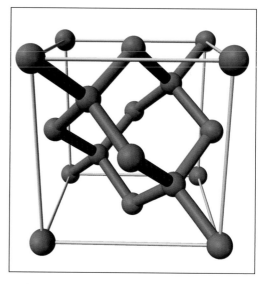

ABOVE LEFT Silicon is one of the most important minerals in Earth and the second most abundant, which, with oxygen, comprises 90% of the crust. *(David Baker)*

ABOVE Silicon crystallisation forms a diamond cubic structure. *(Ben Mills)*

LEFT A ferrosilicon, or an alloy of iron and silicon, can be produced by reducing it with coke in the presence of iron. Intriguingly, a chemical reaction using ferrosilicon, sodium hydroxide and water can be used to quickly produce hydrogen for balloons. *(Focal Point)*

In Earth, most of the minerals originate in the planet's crust, which is 75% silicon and oxygen by element, with silicate minerals comprising 90% of the material from which the crust is formed. There are eight minerals that make up 98% of the crust, which are, in order of abundance, oxygen, silicon, aluminium, iron, magnesium, calcium, sodium and potassium. These minerals are directly controlled by the

LEFT A chemical element within the boron group, aluminium is a silvery-white, non-magnetic metal and is the third most abundant element in Earth's crust. As the most highly used non-ferrous metal it is second only to iron in the quantity used for manufacturing and industrial purposes, all common foils and cans consisting of 92-99% aluminium. *(David Baker)*

bulk chemistry of Earth itself; a magma rich in iron and magnesium will form mafic minerals such as olivine and pyroxene; a silica-rich magma will crystallise to form minerals with more SiO_2, feldspars and quartz.

The electronic structure of the elements is a key to the formation of minerals. In looking at the structure of the atom to understand the way they are determined by atomic mass, the configuration of the electrons occupying successive shells around the nucleus determine the bonding properties that can link them together. As the atomic number (the number of protons) goes up so do the number of electrons and the number of shells to form heavier elements. The hydrogen atom has but one proton and one electron, which resides in the innermost, or K, shell which can accommodate only two electrons. Next up is the L shell, which can hold eight electrons, followed by the M shell with 18, the N shell with 32, the O shell with 50, the P shell with 72 and the Q shell with 98.

The elements are divided into groups according to their electron structure. The inert, or noble, gases such as neon, argon, krypton, zenon and radon have no tendency to form chemical bonds because their outer electron shells are full and stable. The next group up includes alkali metals such as sodium, potassium, calcium and magnesium and these have one or two electrons, called valence electrons, which tend to stray and form cations (an ion with a positive electric charge) to assume the stable configuration of the inert gases. They are the most abundant of the common elements.

The third group are known as transition elements and are the iron, copper and zinc group, whose outer shells tend to be gained or lost from an inner shell rather than the outer one. In many elements of higher atomic number the outer shells are not completely full uniformly and may have their outer electrons in the P shell while gaining or losing electrons from the O

shell to form ions. These can, for example, form different types of ions. Iron forms two cations, Fe^{2+} and Fe^{3+}. A fourth group with valence electrons in the third from outermost shell includes the rare earth and actinide elements, of minor abundance in Earth and not a major player in the story of the planet's structure.

Crystallisation is an important part of understanding mineral properties and is a definition of the chemical bonding of one element to another. Simple connections, such as ionic bonds, form with the electrostatic attraction of ions of opposite charge. The simplest form of ionic bonding is sodium chloride (NaCl), famous in chemistry lessons

BELOW An aluminium bar displaying a fine surface etch under microscopic magnification.
(David Baker)

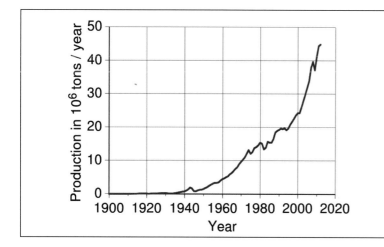

ABOVE The world production of aluminium is driven by commercial and industrial demand. Aluminium is used in major structures in road and rail vehicles, ships and aircraft. *(David Baker)*

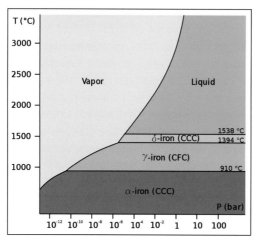

ABOVE An iron meteorite from the core of a planetismal displaying the Widmanstatten pattern of nickel-iron crystals, an interleaving of kamacite and taenite bands commonly known as lamellae. Iron meteorites form only 6% of known falls of all meteorites. *(Waifer X)*

LEFT This diagram shows the low pressure phase for pure iron by temperature (vertical) and pressure (horizontal) where one bar is approximately equal to Earth's sea-level pressure. *(David Baker)*

RIGHT Popularly known as "fool's gold" due to its pale brass-yellow hue, iron pyrite is an iron sulphide. Deriving its name from its use for creating fire, it is usually found with other oxides in quartz veins in either sedimentary or metamorphic rock. *(J J Harrison)*

for being the first bonding process worked out. The ratio of cations to anions determines the stability of ionic bonds which is predictable by the coordination number, the number of equally matched electrons of opposite charge. In the case of sodium chloride there are six cations and anions of equal number in what is referred to as an octahedron coordination.

Although there is no molecule of NaCl, there is a unit cell comprising a geometric form outlined by the minimum number of ions necessary to show the basic building block and which, if repeated a large number of times, will form a visible sodium chloride crystal. This particular unit cell and those of other chlorides are cubic because of the symmetrical spacing of ions. This information, once cracked, enabled scientists to deduce the structured packaging of spherical ions of different sizes. When the ionic radius is known the exact size of the ion can be calculated and trends can be predicted in elements of the same chemical group. The ionic radii of elements in the same chemical group increase with atomic number in relation to the addition of electrons and electron shells, although this in not exactly proportionate because the electrons are added irregularly.

ABOVE LEFT Iron pyrite cube crystals from Navajun in Spain. In the early days of radio, iron pyrite was used as a detector in receivers and was only supplanted when the vacuum tube was invented. As a semiconductor it has many applications. *(David Baker)*

ABOVE One of five members of the alkaline metals group in the periodic table, magnesium is produced in old stars. Significant quantities are present in the Earth's mantle and it comprises 13% of the crust *(CSIRO)*

LEFT Magnesium is used in a wide range of commercial and industrial products, despite its propensity to ignite and burn with a hot flame. It and its alloys are explosive hazards and during the burning process it produces ultraviolet light which is harmful to the eyes. *(ICI)*

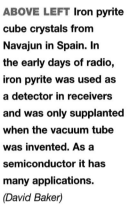

Compounds that achieve a stable electronic configuration do so by sharing electrons rather than losing or gaining them, through what are known as covalent bonds. These depend on the number and distribution of the shared electrons in outer shells, which in turn determine the compounds and crystal structures through more complex factors than ionic bonds. The classic example of covalent bonding is the diamond, where each carbon atom (which is not an ion) is surrounded by four others (with a coordination number 4) in a regular tetrahedron – a polyhedron with four triangular faces, three of which meet at each corner.

For our understanding of the way Earth works it is necessary to consider only a small number of the common minerals abundant throughout the universe. The two most abundant minerals in Earth's crust are oxygen and silicon, which can be divided into a few major structural types. The basis for all silicate structures is the radius ratio of silicon to oxygen, which is 0.30. The radius ratio is the ratio of cation radius to anion radius, and for some structures the cations are too large to fit into the interstices and the anions are held slightly apart (an anion is an atom which has gained an electron).

In some structures if a slightly smaller cation were to be inserted the anions would be held less far apart. This closer-packed structure would be more stable because the electrostatic attraction holding it together would be stronger. If a cation is small enough to fit precisely within the spaces between the anions it would have the most stable configuration, because in such a configuration the anions touch each other. It cannot be stronger than this because even if the cations are smaller still it will have no effect on the anions, since they cannot get closer than contact. In this way it is possible to determine what kinds of structures are possible with which elements and to understand why minerals form the crystals they do.

Making rocks

Scientists and geologists classify minerals through their physical and chemical properties, defined as above through the way atoms and ions are bound together. But this simplified description is only a basic explanation. In reality, bonds are rarely of one type and most are hybrids, and for this reason the correlation between that and their properties is only a broad generalisation. In addition to the chemical types, defined above as the atomic or ionic bonds both possible and plausible, the physical properties are measured in a variety of ways, determining hardness, cleavage, fracture, streak, lustre, specific gravity and density.

Hardness is measured against the Mohs scale, and is the product of the strength of the bonds between ions or atoms. These strengths can vary along the crystallographic axes and so may vary slightly in different directions. It is a partially subjective measure on a scale of 1 to 10, based on the ability of one natural sample to scratch another and leave a visible track, starting with talc ($Mg_3Si_4O_{10}(OH)_2$) as 1 and diamond (C) as 10. Intermediate levels of hardness start at 0.2–0.3 for caesium and rubidium, 0.5–06 for lithium, sodium and potassium before talc (1) with the scale number >10 beyond diamond. Each whole number also has intermediate scales.

Cleavage is defined as the defined planar surfaces along which minerals break, the number varying among minerals. Cleavage planes have some of the same characteristics as their crystal faces and they always occupy definite positions with respect to the symmetry or the crystallographic axes of the crystals themselves, always parallel to a possible crystal face.

The measurement of cleavage is on a scale of poor to good, depending on the type of bonding. Some bonds may be very weak, so-called van der Waals bonds that are a weak electrical attraction related to the asymmetry of certain atoms and ions. Mica breaks very easily while quartz has no cleavage because it is so strongly bonded. In the commercial world, expert cleavage is a highly sought after skill within the diamond industry, cutters in Amsterdam, Holland, being among the most famous in the world for cleaving diamonds. Their expertise can dramatically increase the value of a cut stone.

Fracture is another measure of breaking a mineral other than along cleavage planes and is the most difficult to predict or to work with because it defies simple predictions of how breaks will occur based on a wide range of

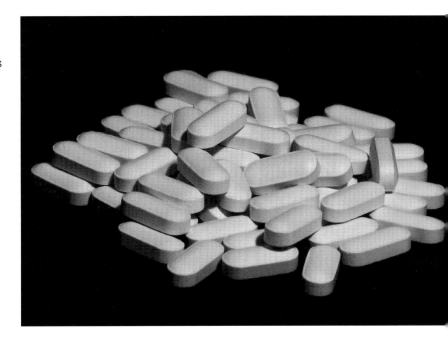

ABOVE Calcium is a vital component of life, a mineral with great health benefits for building strong bones. Bone density depends on adequate levels of calcium, which can degrade in the process of ageing and is one of the early debilitating effects of long duration space flight. *(Ragesoss)*

LEFT The sixth most abundant element in Earth's crust, sodium is found in feldspars, sodalite and rock salt, with sodium ions having been leached from the oceans and seas on the planet over the last few billions of years. *(Denis s. k.)*

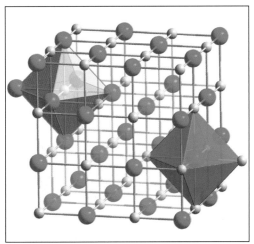

LEFT Sodium chloride (NaCl) structure showing octahedral coordination. *(David Baker)*

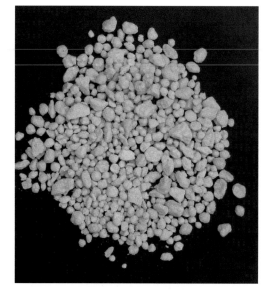

continuously changing variables. In scientific terms it is the breaking of bonds in directions that cut across crystallographic planes. Streak is the colour of the fine powder that is released when a mineral is scraped across a tile of unglazed porcelain, called a streak plate. The colour is the product of the mineral, hematite for instance giving off a reddish brown because it is an iron oxide, regardless of the colour of the specific mineral aggregate (see later).

The lustre of a mineral is a measure of the light reflected off a mineral and is controlled by the index of refraction and the absorption properties of the elements within the mineral. This is often used to provide a quick guide to the identification of certain minerals within a thin slice through which a light has been shone. Most of the attention on lustre is of value only to the precious stones market, but the way light plays on the surface – or through it, to produce stunning colour variations – is a marketable aspect of that business. In scientific terms, lustre can be of value when impurities are displayed in microscopes, such as dispersed flakes of hematite in a quartz crystal, which tells something of the history of that sample.

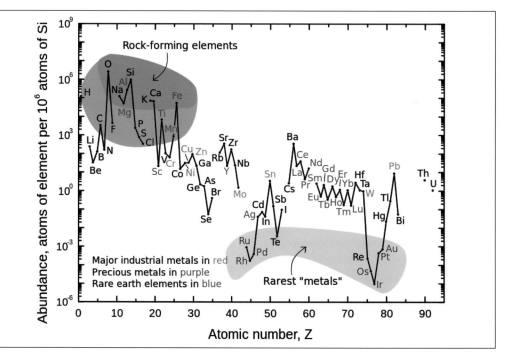

Specific gravity and density are important measurable aspects of certain materials and a scale based on specific gravity has been adopted as a measure of density. Specific gravity is the weight of the mineral in air divided by the weight of an equal volume of water at 4°C (39.2°F). Density is the product of the atomic weight of the constituents and the tightness of the packing of the atoms in the crystal structure and is useful in identifying minerals. For this reason, density has been accurately measured for most minerals.

Minerals are distinguished as being of either primary or secondary origin. Primary minerals are formed by solidifying magma – a product of heating, crystallising in an aqueous solution; by sedimentary processes or extreme pressure; or by metamorphism. Secondary minerals are formed from alteration of the original material through either oxidation or reduction under low temperature or pressure close to the surface of Earth. The definition of minerals helps add to an understanding of the rocks that formed Earth in its various stages of evolution and to classification of type and form, the initial major categories being igneous, metamorphic and sedimentary.

Igneous rocks get their name from *ignis*, the Latin word for 'fire', and form from the crystallisation of a magma, a mass of melted rock forming deep in Earth's mantle. There are more than 700 different types of igneous rock and most form below the crust where temperatures are in excess of 700°C (1,290°F). Extrusive igneous rocks form from rapidly cooling magma erupting from volcanic vents at the surface. Rocks of this type, such as basalt, are recognised by their glassy or fine-grained texture with tiny crystals that have no time to form and grow gradually, since they cool rapidly on exposure to the air.

Intrusive igneous rocks form when the magma intrudes into unmelted rock masses deep in the crust. Here, large crystals have the chance to form because the magma cools slowly, and intrusive rocks can be recognised by their interlocking large crystals that have time to grow during the process of cooling. One such intrusive igneous rock is granite, common in the intrusive igneous rocks that are the pillars upon which large mountain

ABOVE Cleavage is one of several means of characterising the physical properties of a mineral, the others being fracture, streak, lustre, specific gravity and density. This green fluorite shows prominent cleavage planes, vital indicators of how it can be split without shattering. *(Eurico Zimbres)*

BELOW The Hope Diamond, housed in the National Gem and Mineral Collection at the National Natural History Museum, Washington, DC. Diamond is a metastable allotrope (a property where it can exist in two or more different forms) of carbon, its atoms arranged in facet-centred cubic crystal structures. *(David Baker)*

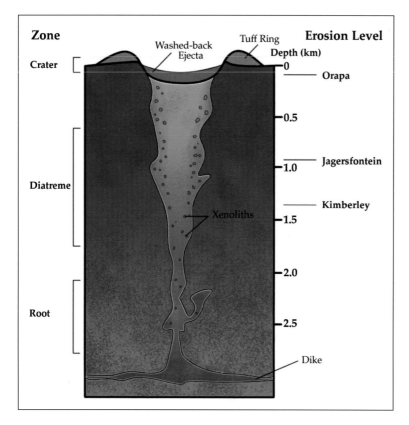

Zone   **Erosion Level**

Washed-back Ejecta   Tuff Ring

Crater

Depth (km)

0 — Orapa

0.5

Diatreme

1.0 — Jagersfontein

Xenoliths   Kimberley

1.5

2.0

Root

2.5

Dike

ABOVE Diamonds are formed in the mantle more than 93 miles (150km) below the surface. They can be brought up in volcanic pipes of diamond bearing rock, which cooled before it could be thrown out. *(Premium Diamonds)*

BELOW Because of their hardness factor, diamonds are used for industrial purposes, as in this application where a pyramidal diamond is attached to the work surface of a Vickers hardness tester, first applied in the early 1920s. *(Vickers)*

chains are built. Very coarse-grained intrusive igneous rocks which form at great depth are known as abyssal, whereas similar rocks which form closer to the surface are known as hyperabyssal.

The igneous rocks are particularly important because they tell scientists a great deal about the composition of the mantle where they originated and from where they are transported to the surface. Their record, in either extrusive or intrusive form, explains a lot about the temperatures and pressures they have experienced at great depth. Certain types of dating can be used to obtain their absolute age and this can be set against geographical strata that then provides a timeline of events. Examining these rocks is a forensic science akin to finding out whether the butler did it! By taking apart the recorded history of the rocks, scientists can reconstruct events that would otherwise be lost.

Metamorphic rocks get their name from the Greek words for 'change' (*meta*) and 'form' (*morphe*) and relate to rocks that are produced when high temperatures and pressures deep inside Earth cause any kind of rock to change its mineralogy, texture or chemical composition while retaining a solid form. The temperatures for metamorphosis are lower than the melting point of rocks (about 1,290°F (700°C) but high enough, above 480°F (250°C), for the rocks to change by recrystallisation and chemical reactions where pressures are in excess of 1,000 bar. Geologists identify metamorphism in two paths: regional and contact.

Regional metamorphism occurs where the high pressures and temperatures extend over very large areas and where mountain building takes place resulting in folds and breaking of layers once horizontal. Where such episodes occur in small areas and where igneous intrusion takes place they are referred to as contact metamorphism. Because their parent rocks are rich in silicates, these are the most abundant minerals in metamorphic rocks; typical types are quartz, feldspar, mica, pyroxene and amphibole. These are also the same kind of silicates found in igneous rocks.

By volume, igneous and metamorphic rocks

comprise 95% of Earth's crust to a depth of
10 miles (16km) and together they tell a story
about how the rocks were formed, when and
under what circumstances. Metamorphic rocks
alone account for 27.4% of Earth's crust. They
provide the bulk evidence for the long-range
history of our Earth and of its potential future
course, but there are other materials left around
as a product of violent short-term, and passive
long-term, episodes that take the igneous and
metamorphic rock and rework them into more
discrete episodes, leaving their record behind
across much shorter timescales. These are the
sedimentary rocks.

Sedimentary rocks comprise the
accumulation of fragments, separated or
cemented together, that include rocks, minerals,
organic matter and substances formed in
aqueous solutions – water. Organic particles
are known as detritus, pieces of shattered
rock are known as clastic sediments, and
these accumulate in places where minerals
can chemically precipitate from a solution. This
process leads to compaction and cementation
under exposure to moderate pressure and
temperatures. Here too the past history is
revealed through classification. Detritus consists

of minerals broken or fragmented from pre-
existing rocks and is often moved by aqueous
currents or by glacial flow. The rocks from
ancient mountains worn down by erosion and
weathering are clastic and can be reconstructed
from their detritus.

Because of this active transport mechanism
in fluids or by shifting ice, the fragments
become eroded, weathered and rounded. In

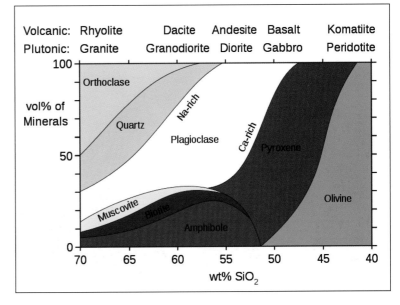

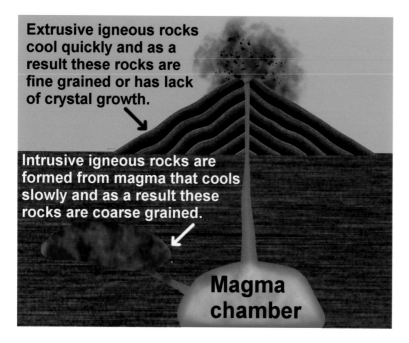

Extrusive igneous rocks cool quickly and as a result these rocks are fine grained or has lack of crystal growth.

Intrusive igneous rocks are formed from magma that cools slowly and as a result these rocks are coarse grained.

Magma chamber

ABOVE Igneous rocks can be formed intrusively in magma chambers surrounded by existing rock and extruded as volcanic rock through fissures or from eruptions where they cool rapidly on contact with air. *(Jasmin Ross)*

the process of sedimentation, currents sort out these materials by size and weight with varying efficiency, and thus form the basis for subdividing them into coarse-grained gravels and hardened equivalents, medium-grained sands and sandstones, and fine-grained clays and muds, and their lithified equivalents, shales. Coarse sedimentary rocks formed from angular gravel cemented together in a

matrix are known as breccias, as against the rounded and smoothed pebbles and cobbles of conglomerates.

A word of caution. In addition to being formed from consolidated material accumulated from slopes and cliffs, the term 'breccias' can also be applied to agglomerates thrown out of volcanic vents as blocks of lava in an ash matrix, and here they are the product of explosive eruption. They can also be formed in hydrothermal fractures where an existing rock is shattered by thermal or pressure-induced stresses and several separate rocks are cemented together in a binding cast cemented tight by calcite, silica or iron oxide. Lunar geologists are used to speaking of Moon rocks as breccias, and these represent consolidated agglomerates welded into a single rock by eruption of volcanic activity either in flow or by extrusion.

Sedimentary rocks are sometimes precipitated from solution and their minerals can identify the parent solution. The most abundant chemical rocks are limestone and dolomite, which is largely made up of calcium and magnesium carbonate. Limestones are largely made up from calcareous fossils, shells formed by the biochemical precipitation of calcium carbonate that animals extract from water. There are some famous limestone landmarks, not least the White Cliffs of Dover and their opposites on the French coast, reminiscent of the time before the English Channel broke through, linking the North Sea with the Atlantic Ocean.

Together, shale, sandstone and limestone comprise more than 95% of the total sedimentary component of Earth's crust, of which shale accounts for around 70%, sandstone about 20% and limestone 10%. However, placed in perspective, sedimentary deposits account for a mere 5% of Earth's crust, the balance being in igneous and metamorphic rocks as described above. Collectively, these three categories define

2.0 cm

LEFT Gabbro is a dark, course-grained, intrusive igneous rock which forms when molten magma cools over time. It forms the base upon which the oceanic crust rests. This specimen is from Rock Creek Canyon, Sierra Nevada, Calif. *(David Baker)*

the makeup of the outer layer of Earth, the lithosphere, which is defined as the upper rigid part of the mantle and lower crust. But whereas the whole Earth by abundance is dominated by iron (35%), oxygen (30%) and silicon (15%), the crust alone is dominated by oxygen (46%) and silicon (28%).

Now it is time to pick up the story of Earth's evolution, beginning around 4 billion years ago at the end of the Hadean eon and the base of the Archean.

Forming Earth

The Earth machine can be divided into separate layers from the centre up through the outer surface and above into the atmosphere. So far we have seen how Earth formed and from what. We have learned that the Moon played a pivotal role in changing the nature of the planet and in adding about 10% to its mass from the remains of the small planet Theia. Now it is time to see Earth in the context of a series of integrated systems within the machine. These are the geodynamic system, the tectonic system and the climate system.

The geodynamic system is the product of activity in the inner core and the outer core. The tectonic system that keeps Earth geologically active and gives it an ever-changing surface is controlled in the deep mantle, in the asthenosphere (the lower part of the upper mantle) and the lithosphere (the crust), which comprises Earth from the bottom of the mantle to the surface. The climate system is the responsibility of the hydrosphere (the seas and oceans) and the atmosphere (the air above the surface), which together provide weather.

Recently, scientists have been considering whether there is a fourth geologically relevant layer – the anthroposphere, which relates to humans and the impact they have had on Earth, as the only animal known to have so significantly altered the planet that it has been deflected to a different evolutionary track from that which it was on before interventions through large-scale agriculture and industrialisation. As we will see in Chapter 7, the evolution of Earth's surficial order is itself a product of life, which has structurally remodelled the atmosphere to such an extent that it has

dramatically altered the climate of the planet for at least 250 million years. So perhaps we are not so special after all.

In Chapter 2 we explored the origin of Earth up to the end of the Hadean period 4 billion years ago and we will now pick up that evolution and take it into the Archean, a period spanning 1.5 billion years. It is a period during which the development of Earth settled into the changing cycles we live with today; but in studying the Archean we come upon some surprising discoveries that have changed the way scientists interpret the early Earth.

It used to be thought that the planet was molten for several hundred millions of years. It was this which, it was said, best explained the scarcity of rocks whose radiological clocks were reset around 4 billion years ago. When geologists refer to the 'age' of a rock it is usually a reference to the last major event that changed it to the form in which it is analysed today. The same materials from which all the features on Earth are explored today were present when Earth accreted from debris and the protosolar nebula when the solar system began. In this context, the age of the rock indicates the most recent time it was thermally disturbed to reset its clock through metamorphism or crystallisation. But more of that later.

It is now believed that Earth had a surface crust and an ocean (see Chapter 5) during the Hadean eon as indicated by the zircon rocks found in Australia. The isotopic composition of

BELOW Granite is an intrusive igneous rock, beautifully displayed in this example from Chennai, India. *(Mark A Wilson, USGS)*

oxygen in these crystals ($^{18}O/^{16}O$) confirms the existence of liquid water 4.4 billion years ago. This is not consistent with a picture previously drawn showing Earth having a magma ocean across its entire surface. The oldest rocks discovered to date are the Acasta gneisses dated to 4.031 billion years.

The chemical formula for zircon is $ZrSiO_4$, and it is widely found in iron-rich magmatic and metamorphic rocks. It is a very hard mineral with a Mohs scale factor of 7–7.5 and is therefore resistant to erosion and alteration (a geologists' words for any form of change). When it crystallises the zircon lattice incorporates uranium and thorium. These radioactive isotopes (^{238}U, ^{235}U and ^{232}Th) decay into lead over a period of time longer than that of the solar system. Because of this the pairing of uranium and thorium serves as an accurate age-dating clock.

This highly accurate and reliable method is the one used to determine the age of these zircon crystals at 4.404 billion years, but the age of these detrital remnants does not age the rock in which they are found – samples reworked much later, acting as a host and into which the crystals have been inserted. In most cases

the adopted parent rock has gone, through erosion or fractionation, leaving the crystals to be carried by later processes into the structures of other rocks. In this way, in some cases, it is possible to trace the history of the crystal. But there is another indicator that adds to the forensic analysis to find out whether the butler really did do it!

The smoking gun for an early separation of mantle and crust, and a cooler, water-bearing surface, comes via another forensic route. Zircon is one of the first minerals to crystallise at high temperature in acidic magma. These acid magmas are rich in silicon and poor in magnesium, which is the opposite to basic magmas. Silicon-rich acid magmas result in the crystallisation of quartz (SiO_2), and where this takes place at depth these rocks – known as plutoids – are referred to as granitoids, rocks which include granite, granodiorite, etc, all containing quartz but differing from each other in the relative abundances of feldspar and its alkaline varieties.

From this evidence alone there is proof that the granites already existed 4.4 billion years ago, water being the essential component in their formation. As the zircon crystals grew they incorporated some of the minerals that crystallised in the magma at the same time, and the Jack Hills zircons do contain inclusions of quartz, feldspar, plagioclase and potassium along with several other minerals. All these are characteristic of granite, and nail the zircon crystallisation in an acid magma. How wonderful it would be if there was physical evidence of the existence of a piece of Earth's crust from this astonishingly early period. Well, there is!

The zircon crystals from Jack Hills contain rare earth elements around in the magma when they crystallised. Rare earth elements are some of the most abundant in the crust and are called 'rare' due to their scarcity in ore deposits where commercial mining takes place. Apart from scandium they are all heavier than iron and were produced by nucleosynthesis in supernova explosions. All 17 chemical rare earth elements are divided into two families: those that have lost all their magnetic characteristics (see later in this chapter); and those that have lost them due to late hydrothermal processes.

The Jack Hills zircons have the same rare earth elements as those from the Acasta gneisses, which crystallised from magmas typical of those found in the Archean crust and distributed between 4 billion and 2.5 billion years ago. Based on the known distribution of rare earth elements, it is possible to calculate the rare-earth content of the magma in which they formed. It turns out that the host magma for these zircons was abundant in light, and depleted in heavy, rare earth elements. This is a typical fingerprint for magmas that produced the Archean crust.

A lengthy analysis of the Jack Hills zircons and a wide range of other detrital materials enables scientists to build a credible picture of the growth and recycling of the early crust, which appears to have occurred in peaks around 4.4 billion and 4.3 billion years ago, and

again between 4.2 billion and 4.0 billion years. The origin and forming of the crust we know today was not completed in a single step, but all of these sequences imply a cool surface with water in several oceans across the planet. But this is far from the world we know today and to discover that we have to start with the Archean period.

When looking back at the early history of the Moon, the Late Heavy Bombardment was considered to be a particularly important time for both our natural neighbour and Earth (see Chapter 2). The LHB provided a reworking of the surface of the Moon, but these events that are so visible today also occurred on Earth, which also experienced a rain of rocks and planetismals that not only reworked the surface but probably introduced many of the essential elementary materials for life itself. Those episodes, so evident in the geologic record right across the lunar surface, are indicative of the early history of Earth; and, as we saw earlier, all the terrestrial planets bear

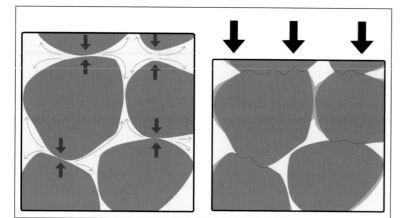

ABOVE Sedimentary materials can form under pressure as clastic rock (several pre-formed rock fragments or minerals), dissolving when the grains are in contact (right) or crystallising when there are vacant pores. There is always a flow of material from higher to lower stress levels. In this way loose sand forms sandstone. *(Woudloper)*

evidence of early bombardment right across the solar system.

We learned earlier that there are about 300,000 craters with diameters exceeding 0.6 mile (1km) on the front face of the Moon alone. Most of those were formed in the first billion years, but of that total only 1,700 craters are greater than 12 miles (20km) in diameter while only 15 have diameters in the size range 186–745 miles (300–1,200km). These are in the Imbrian and Nectarian periods and give a useful analogue for calculating the number of incoming planetismals to which Earth would have been subjected, largely between 3.95 billion and 3.87 billion years ago.

With Earth's larger mass and stronger gravity field the incoming flux would have been between 13 and 500 times greater than that of the Moon. Accordingly, estimates conclude that approximately 22,000 planetismals struck Earth leaving craters greater than 12 miles (20km) in diameter, with between 40 and 200 craters having a diameter greater than 620 miles (1,000km). Some may even have been the size of continents, with diameters of up to 3,100 miles (5,000km). While there is little or no physical evidence to show where these impacts occurred – most traces would have been wiped out by the movement of the crustal plates and continental materials through tectonic activity – the fingerprints of the guilty have been found.

The Archean sedimentary record contains unexpectedly high levels of platinum-group elements, high abundance of shocked minerals and also strange isotopic combinations. The platinum group is a family of rare elements consisting of ruthenium (Ru), rhodium (Rh), palladium (Pd), osmium (Os), iridium (Ir), platinum (Pt) and occasionally some rhenium (Re). These platinoids, as they are called, are relatively common in meteorites but very rare in the mantle, where Ir is only about 3.2ppb (parts per billion) or the crust (0.03ppb). Also

RIGHT Sedimentary rock such as sandstone can be seen in many of the Mediterranean islands, including Malta where this example is found. *(David Baker)*

found are tektites, formerly molten silicates with a glassy texture from increased temperature at impact, with shocked minerals, formed literally by deformation under very intense and sudden pressure from shock waves.

A critical piece of evidence for the overwhelming bombardment suffered by Earth is in the absence of rocks from this particular period between 3.95 billion and 3.87 billion years ago, a relatively brief 80 million years. Anything of that size on the moon could easily have been experienced by Earth. But there is more evidence from the isotopes. Only since the early part of this century have analytical techniques become available to allow the detection of isotopic variations in tungsten (W), the ratios of $^{182}W/^{183}W$. These were measured in rocks from Greenland and Labrador and the measured ratio set against a normalised ratio for the average terrestrial value versus the value from meteorites such as chondrites.

The normalised value for Earth (ε_w) was set at zero, with the post-Hadean rocks showing $\varepsilon_w 0$-+1 while meteorites show between ε_w-1 and -2.5. Samples of tungsten isotopes from the Greenland and Labrador regions showed a range of from ε_w-0.5 to -2. If the measured samples were all of terrestrial origin they would have shown around zero to +1 instead of the meteorite ratios between -1 and -2.5. There is clearly a bias toward meteoritic origin with a proportionate drift showing a heavy influx of meteoritic material mixed with the terrestrial materials. But there are opponents of this theory who have searched unsuccessfully for a similar imbalance in chromium isotope ($^{53}Cr/^{52}Cr$) ratios. But here perhaps it is best to remind ourselves that absence of evidence is not evidence of absence.

Within the data too is a partial solution to the question over whether there was a continuous period of bombardment of Earth from the time of its formation to 3.85 billion years ago. A period of 600 million years in which, had Earth been subjected to the same conditions as some scientists thought existed on the Moon, the entire surface area would have remained a boiling magma ocean – which we know, from aforementioned evidence, it was not. So there must have been a series of episodic events, allowing partial recovery of the surface back to less violent conditions in between surges so that the total quantity of meteoritic material falling down on Earth was spaced over intervals.

Yet for some time after the golden era of Moon sample-gathering in the 1970s, scientists

BELOW Stratigraphy is a way of determining a sequence in which things happened, upper layers superposed over earlier deposits indicating they came later. This example from the Colorado Plateau of southeastern Utah displays (from the top) rounded domes of the Navajo Sandstone, layered red banding, vertically jointed red sandstone, purple Chinle Formation, lighter red formation, and white Cutler Formation sandstone. A vertical geological history of the Permian to the Jurassic. (David Baker)

RIGHT Sedimentary deposits on Mars, where the NASA Curiosity roving vehicle landed inside Gale Crater to explore the deposits and discover what kind of climate existed on Mars in the past. In this view several sedimentary layers can be seen, indicative of an earlier period when water flowed freely across the surface. *(NASA)*

Point Lake outcrop

Gillespie Lake sandstone

Sheepbed mudstone

LEFT Marble is an example of metamorphic rock, this one from the Mississippian geologic region in Big Cottonwood Canyon, Wasatch Mountains, Utah. *(Mark A Wilson)*

believed otherwise, imagining that as the big basin impacts and major cratering epochs appeared to have been more extensive further back in time it was a continuous process but gradually declining. Cratering rates on the Moon show a sharp increase the further back in time one goes. From this, it was said there was an unbroken phase from the origin of the Earth–Moon system when intense bombardment was at its most intensive at the beginning, declining gradually over the first 600 million years of the solar system. But if that had been the case the bombardment was so intense that the Moon itself would have formed as late as 4.1 billion years ago, because the amount of material falling in would have equalled the mass of the Moon itself, rather than before 4.5 billion years which we have proven it to be.

As well as this, if the intense bombardment

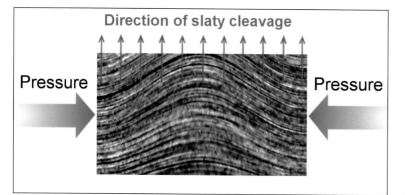

Direction of slaty cleavage

Pressure

Pressure

LEFT An example of dynamic metamorphism where folds and pressure-induced deformation can produce foliation which gives an indication of cleavage. *(David Baker)*

had in reality occurred unbroken over a period of 600 million years, the intense meteorite activity should have enriched these two worlds with siderophile elements, such as the platinoids. There is no such enrichment and there are no tektites found older than 3.92 billion years, which firmly fixes the intense bombardment as beginning around this time. Added to the zircon data and other evidence for an early, and recycling crust, a picture emerges of a relatively quiescent planet between the very early phase of its accretion and the beginning of the late bombardment – there never was a linear and declining intensity continuously stoking the magma fires of a global ocean of molten rock.

The inevitable conclusion is that the heavy cratering cannot be extrapolated back in time because nothing fits. To have that degree of inundation would have created an object the size of the Moon in just 100 million years. But this leaves unanswered the question as to why there should be such an intense, and relatively confined, period of such intense activity. To find that answer it is necessary to go back to the outer solar system and import the latest findings regarding the overall geometry of the solar system itself. At the end of Chapter 1 it was said that the planets did not form in the places they occupy today. This is a key to understanding the reason for the commotion that erupted to cause the bombardment of Earth and Moon just under 4 billion years ago.

Importing several mathematical models from the refined orbital measurements of the planets allows astronomers to calculate their ephemerides for several centuries, sometimes several thousands of years, into the future. Through planetary exploration using unmanned spacecraft, the exact location of the planets has been determined to within a few miles at most, and to fractions of a mile for some. Reverse extrapolation allows scientists to adjust

the positions of the planets as they interact or have their orbits interfered with through interaction with each other and/or the Sun. This allows positional mapping back through time and to computerised simulation of what their movements were many hundreds of millions of years ago. This is not a precise science because even in the space age, the positional precision with which each planet sits within a sequence of astronomical coordinates is not that accurate to eliminate every conceivable error. But there is a universal indication that it was the outward migration of an early, and more compact, solar system that caused the heavy bombardment.

The simulations show that the gaseous outer giants of the solar system were closer in than they are today, occupying a region around the Sun between 5.5 and 15AU (1 astronomical unit

RIGHT Formed by metamorphism from an original formation of igneous or sedimentary processes, gneiss is frequently formed from layers of planar structures, or foliation. This example from the Czech Republic shows banding created under high pressure and temperature. *(David Baker)*

BELOW Contact metamorphism indicates where a rock is altered by igneous intrusion and can become more coarsely crystalline. *(David Baker)*

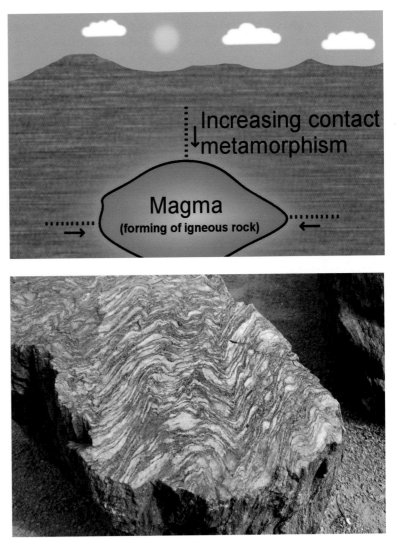

being the distance between Earth and the Sun) rather than the spacing between 5.5 and 30AU that they occupy today. In this early phase of the solar system, within the first few hundred million years, Neptune was in an orbital path between Saturn and Uranus rather than being the outermost planet as it is now.

At this very early date it appears that the icy mini-worlds of the Kuiper Belt were densely packed in a broad swath around the solar

system between about 18 and 35AU. The total amount of material in this region was equal to 35 Earth masses, and interaction with the closely spaced outer planets shifted the Kuiper Belt so that it began to seriously perturb the outer giants – so much so that Jupiter began to migrate inward while Saturn, Neptune and Uranus migrated outward. Computer simulations show that the giants Jupiter and Saturn achieved an orbital resonance in their periods so that Jupiter went around the Sun in exactly twice the time it takes for Saturn, a 2:1 ratio which, but for other gravitational forces, would have been stable.

However, quite quickly after Jupiter migrated inward and Saturn moved outward, all four outer giants were destabilised by this gravitational ballet to the extent that their orbits became eccentric, with Neptune moving into an elliptical path carrying it at its farthest point beyond the orbit of Uranus and into the inner region of the Kuiper Belt. The disruption this caused pulled and tugged at the Kuiper Belt objects and flung them into highly elliptical paths that sent showers of them hurtling toward the inner solar system. Calculations show that approximately one-millionth of the total number of Kuiper Belt objects were gravitationally attracted to the Earth–Moon system, of which 10% would have struck the Moon.

Vast numbers of Kuiper Belt objects would have peppered the solar system with rocky fragments, showering the planets and moons of the outer giants; some of those now natural satellites probably captured Kuiper Belt mini-worlds, pulverising the terrestrial worlds of the inner solar system. The dramatic perturbations caused by this game of gravitational musical chairs shifted Neptune into a more nearly circular orbit beyond the orbital radius of Uranus and, as the vast mass of the Kuiper Belt spent itself on impacting the eight planets and their moons, the two outer giants slowly migrated to their present positions and shifted the inner edge of what remained of the Kuiper Belt out to 30AU where it remains today. One of the larger Kuiper worlds, the bi-planetary Pluto–Charon system, remained in a perturbed orbit so that at its closest approach it passes just inside the orbital radius of Neptune once every 248 years as it slowly orbits the Sun.

After this triggered bombardment of the planets caused by the migration of the outer giants – worlds that in themselves may have insulated the four terrestrial planets from an even more intense assault – the mass movement of the Kuiper worlds and the gravitational movement of the outer giants unleashed what had previously been a relatively stable Asteroid Belt. Of the total number of bodies in the Kuiper Belt only one-thousandth remained after this disruption, but the asteroids themselves were thrown around in a giant game of cosmic billiards, smashing into the planets on unpredictable trajectories. Estimates project that 90% of the original mass of the Asteroid Belt could have crossed the Earth–Moon orbit, feeding into the damage done by the Kuiper objects. Evidence for this, while obliterated on Earth, can be found on the Moon in samples collected near major impact craters exhibiting a chemistry identical to that of chondrites.

The elegance of this new understanding lies both in its simplicity and its universality, in that it explains so many puzzling aspects of the solar system that no other model has been able to satisfy. It also explains many problems with other theories about Earth and the Moon. This phase in the early Earth was crucial to the remodelling of the planet during the Archean eon which followed. But it is difficult to overestimate the severity of the bombardment, during which several oceans probably boiled away and the entire Earth was reconfigured on the surface and to some considerable depth without violating the chemical order making up the mantle and the crust.

What had been envisaged for the sustained heavy bombardment of older models could be invoked for this brief period, lasting less than 100 million years, of the Late Heavy Bombardment, which we began by describing in its impact on the Moon's history, and now incorporate into the story of Earth. Massive volcanic events would have been triggered by this

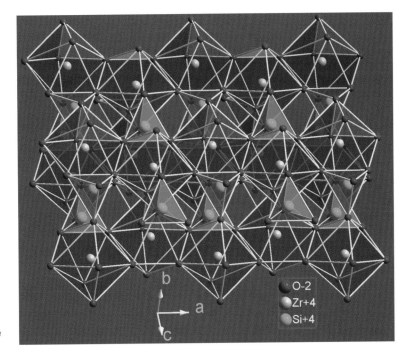

LHB and it is difficult to see how the planet could have continued to survive an episodic event on this magnitude continuing for longer periods of time. While we will discuss the oceans and the atmosphere in later chapters, it is unlikely that all the water on the planet was vaporised.

One of the really big questions is whether life originated before this intense bombardment caused so much physical destruction across

ABOVE The crystal structure of zircon showing the tetragonal form. *(David Baker)*

BELOW The essential characteristics of Earth showing the granitic continental crust and the basaltic ocean crust overlying the asthenosphere on the two mantle regions. The internal structure of Earth is no template for other worlds, where unique conditions have produced a wide range of configurations. *(NASA)*

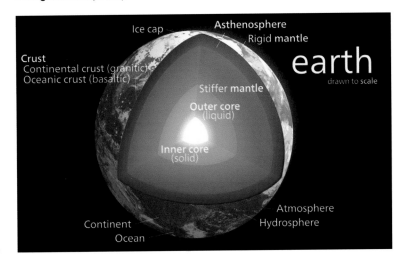

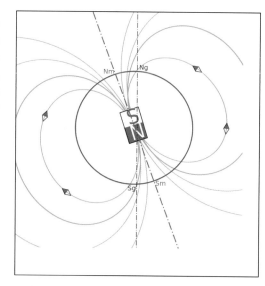

RIGHT The magnetic polar axis is not coincident with the axis of rotation, magnetic north being offset by varying amounts with geographic north. *(David Baker)*

the planet, or after it subsided and the impact events tailed off. We will return to this question in Chapter 6. Notwithstanding almost unanswerable questions of that kind, this chronological dating sequence fits very neatly into the heavy bombardment phase on the Moon which created the great basins, and the direct evidence from lunar samples has gone a long way to pointing scientists toward the

solutions outlined here. But there is one other aspect of the early Earth that we need to import before moving on through the Archean to the world we know today.

Radiation belts

Magnetic fields are found in all terrestrial planets with iron cores, and Earth's magnetic field is especially important for the emergence and preservation of life since it creates an enveloping shield that protects the planet from hard radiation that would otherwise reach the surface. The magnetic field creates what is known as the magnetosphere, and it is this which scientists have found most fascinating to explore since the dawn of the space age. But the very existence of radiation belts surrounding Earth, comprising particles trapped within the magnetic field lines, was not proven until the late 1950s.

It was only with instruments on the first US satellite, Explorer I, launched on 31 January 1958, that earlier theories about the existence of such bands surrounding Earth were verified. Since that

RIGHT James A Van Allen demonstrated that the long-suspected radiation belts surrounding Earth were real when he instrumented America's first satellite, launched by the US Army in January 1958. Here the NASA Lewis Research Center creates a simulated radiation belt around a model of Earth, showing the toroidal shape. *(NASA)*

date numerous scientific satellites have explored this region and mapped the magnetosphere and the zones of trapped radiation which are named after the Explorer I instrument designed by Dr James A. Van Allen of Iowa State University. The inner and outer Van Allen radiation belts sit within the magnetosphere, invisible shells of dangerous particles that protect the lower atmosphere and the surface of Earth, all driven by magnetic forces.

Earth's magnetic field is generated by the motion of conductive fluids and in iron alloys extending out about 2,110 miles (3,400 km) from the centre of the planet. The liquid outer core is powered by thermal energy from the inner core. The inner core has a radius of approximately 760 miles (1,220 km) from the centre of Earth. The temperature of the inner core is about 6,000 K (5,730°C), producing a thermal flow that cools to around 3,800 K (3,530°C) at the boundary with the mantle. It is this inner core and the rotation of Earth on its axis that produce a geodynamic effect, generating a magnetic field through the simple process of a dynamo.

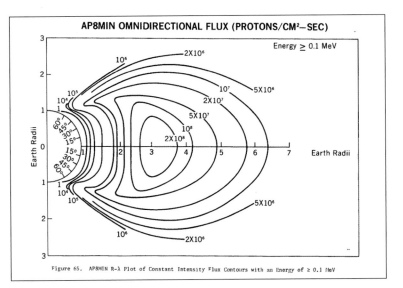

Figure 65. AP8MIN R-λ Plot of Constant Intensity Flux Contours with an Energy of ≥ 0.1 MeV

A common misconception is to believe Earth to have a bar magnet permanently installed in the solid core. This cannot be the case because extreme heat destroys magnetism. The majority of materials lose their magnetism at the Curie point, named after French scientist Pierre Curie, which is a temperature of 773 K (500°C). This temperature is already exceeded at a depth of only 12–20 miles (20–30km) below the surface,

ABOVE Because the Van Allen belts occupy a toroidal, or doughnut-shaped, structure, the Moon-bound Apollo astronauts flew a path up and over the most intense radiation zones thus avoiding the exposure they would have received flying straight out along the equatorial plane. *(NASA)*

LEFT The solar wind is deflected by Earth's magnetosphere, protecting the atmosphere from the destructive effect of high energy particles which in the early stages of Earth's evolution had stripped it of its first atmosphere. *(NASA)*

so Earth cannot be permanently magnetised below this depth. Because Earth's outer core is fluid, it creates what scientists call a self-exciting dynamo, where small, stray magnetic fields from the convective iron produce an electric current that then produces a magnetic field with opposing polarities (north and south).

These opposing polarities induce a feedback process based on a loop. The changeable magnetic field creates an electric field and both this and the magnetic field exert a force on the charge flowing through the currents. The motion of the field is sustained by convection and that motion is driven by buoyancy. As the temperature increases toward the centre of Earth it provides the buoyancy, and as the outer edge of the liquid core cools some of the molten iron solidifies and is plated to the inner core.

While Earth's magnetic field conforms to a magnetic dipole, with its axis close to the planet's rotational axis, sudden and quite dramatic reversals have occurred in which the poles switch places. The rock records of Earth show these magnetic reversals and this information is used to understand the tectonic movement of Earth's continental plates (which

see). As they cool down below the Curie point the polarity is 'frozen' into the rock as a memory of what the polarity was when that cooling took place. If the rock is reheated, it passes again through the Curie point and would cool with the new magnetic polarity embedded in the sample. Several examples of magnetic reversal have been recorded by geophysicists and by oceanographers collecting cores from the sea floor.

The declination of the polarity – a measure of the angle between magnetic polarity and the polar rotation axis of Earth – also changes. A compass needle in London swung from 11° east of true north in 1580 to due north in 1660, to 24° west of north in 1820 and back to 7° west of north in 1970. This is an indication of the fast motion of the liquid outer core, where the magnetic field is generated. This is much faster than any geologic change takes place and is caused by the fast flowing cells of energy in the liquid core. The fluid motion accounts not only for its presence but for its fluctuations.

The intensity of the magnetic field is measured in gauss (G) or sometimes in nanoteslas (nT) where 1G (one Gauss) equals

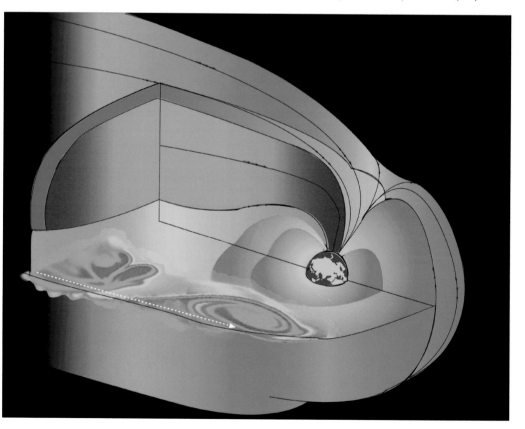

RIGHT The shape of the magnetosphere is dictated by the presence of the radiation belts and by the pressure exerted by the solar wind which, because it various with fluctuations in the Sun's atmosphere, can take on different structural forms. *(NASA)*

10^5nT. Earth's magnetic field measures between 25,000nT and 65,000nT (0.25–0.65G). This intensity level is quite low, a refrigerator magnet having about 100G (0.010nT). The intensity of the field decreases toward the equator and rises again at the poles. The inclination of the magnetic field is given as the angle between 'up' and 'down', where in the northern hemisphere it points down at the north magnetic pole and rotates through 90° as the field loops down across the equator, to point straight up at the south magnetic pole.

The Van Allen radiation belts are layers of energetic charged particles held in place by Earth's magnetic field. Earth has two permanent belts plus transient belts that can appear and disappear naturally or through manmade causes, the latter with unexpected effects. The inner belt typically occupies a region between 600 miles (1,000km) and 3,700 miles (6,000km) containing high concentrations of electrons with several hundred keV (thousand electron volts) and energetic protons exceeding 100MeV (million electron volts) held and trapped by the strong magnetic fields in this region. Because of the slight offset in the declination of the

magnetic field the closest approach of the inner belt to the surface of Earth occurs over the South Atlantic and is known as the South Atlantic Anomaly. In 2014 NASA discovered a ribbed effect showing an oscillation in the field which may be caused by the tilt of the field producing a weak electric field in the inner belt, pulsating through the belt and giving the appearance of zebra stripes.

The outer Van Allen belt consists mostly of high-energy electrons at 0.1–10MeV which extend from about 8,100 miles (13,000km) to approximately 37,300 miles (60,000km). In addition, the outer belt carries various ions (atoms with a disproportionate number of electrons and protons) in the form of energetic protons, and is much larger than the inner belt. It is also prone to separation into two layered forms that eventually remerge to form a single outer torus.

The artificial creation of a radiation belt around Earth was a consequence of Operation Starfish, involving a Thor missile fired from Johnston Island in the Pacific Ocean on 9 July 1962. Detonated at an altitude of 250 miles (400km), the weapon's nuclear

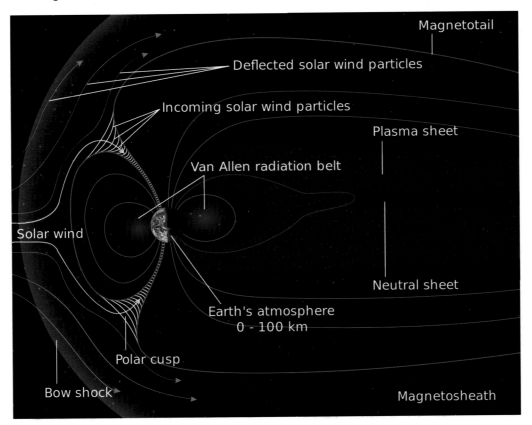

LEFT A diagrammatic representation of the magnetosphere and the radiation belts showing the magnetotail which extends far beyond the distance of the Moon's orbit. *(NASA)*

warhead delivered a yield of 1.4MT (millions of tonnes of TNT equivalent). The electromagnetic pulse (EMP) pushed measuring instruments off the scale, knocked out streetlights and telephone lines in the Hawaiian Islands, triggered burglar alarms and incurred large-scale electrical damage. It also knocked out several satellites in Earth orbit far from the point of detonation, and created an artificial radiation belt that lasted for several days. Subsequent test agreements banned nuclear tests in space, and, eventually, nuclear weapons testing of all kinds under, on, or above the surface of Earth were outlawed.

The Van Allen belts were recognised as a threat to life and a constraint on space flights carried out within the belts for lengthy periods of time. Until the Apollo missions to the Moon, only one spacecraft and two astronauts had ventured, albeit briefly, into the inner Van Allen belt. During the Gemini XI mission launched on 12 September 1966, astronauts Conrad and Gordon used an Agena rocket stage to propel their spacecraft to a maximum height of 850 miles (1,368km). They suffered no ill effects from this experience, due to the very brief few minutes spent in this region.

Until the Apollo missions began flying lunar return flights, no other spacecraft had passed into the Van Allen belts. With the exception of Gemini XI, all space flights in Earth orbit have been conducted below the inner Van Allen Belt. But the flight of Apollo 8 launched on 21 December 1968 required the spacecraft

to pass through this region on its way to the Moon. It was followed by eight more lunar missions, six of which landed on the surface, but on all flights the trajectory selected arched over the top of the inner belt and passed very quickly through the upper, and less energetic layer, of the toroidal outer belt.

Moon-landing deniers have frequently claimed that humans cannot survive the Van Allen Belts and that therefore the flights did not take place, seemingly unaware that the chosen flight path minimised considerably the amount of radiation exposure. Moreover, travelling at more than 20,000mph (32,000kph), the spacecraft rapidly transited this upper region and was beyond the zone of harmful radiation. That said, they received more radiation from solar particles outside the belts, receiving, depending on the mission, between 0.16 and 1.14 rads, or 1.6–11.4mGy where the Gray is the scale measurement. This is less than radiation workers on Earth are allowed to absorb in any year.

One of the more important features of Earth's magnetic field is the magnetosphere, an area of space around Earth resembling a sphere, toroidal in shape, where the Van Allen belts reside, because it ensures the survival of life on Earth by protecting the planet itself from harmful solar radiation. On the dayside of Earth the magnetosphere is greatly compressed by the solar wind (which see), which exerts a pressure on it much as a flowing stream would create a bow wave around a stick placed vertically in the flow of water. On the sunward side of Earth (upstream) the magnetosphere is compressed to a distance of 40,000–56,000 miles (60,000–90,000km) from the surface, depending upon the strength of the solar wind. This is the bow shock front where the solar wind encounters Earth's magnetosphere.

The definition of 'shock wave' is where the majority of the solar plasma drops from supersonic to subsonic speeds, but this does imply the speed of sound as a measurement. The speed of sound in physical terms is defined by an equation taking account of the ratio of specific heating, the pressure and the density of the plasma. The charged particles in the solar wind spiral along the Sun's magnetic field lines, and this gyration is the equivalent of thermal

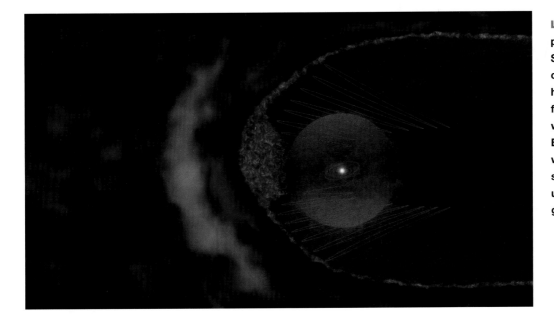

velocity in a gas (such as air), and the mean value is the speed of sound. As the particles encounter Earth's magnetosphere their forward velocity falls below the speed at which the particles are gyrating and this determines the location of the bow shock.

Behind the bow shock is the magnetosheath, which is the region of the shocked solar wind particles and some plasma from the magnetosphere. It is a place with high particle flux with strength and direction fluctuating wildly. In some respects it can be considered as a buffer between the shocked wind and the region below known as the magnetopause and is typically 40,000 miles (60,000km) from Earth on the solar side of the planet, facing directly into the solar wind, which flows around and past Earth leaving a tadpole-shaped tail on the anti-solar side extending to a distance of almost 4 million miles.

Below the magnetosheath lies the magnetopause, a defending barrier between Earth's radiation belts and the solar wind. As the solar wind fluctuates in intensity according to variations in the activity of the Sun, the magnetopause creates ripples and waves where most of the particles from the solar wind are deflected around the planet. This is the line of juxtaposition between Earth's magnetic field and that of the Sun and is similar to the Sun's heliospheric boundary discussed in Chapter 2, and found by transiting spacecraft to lie at about 110AU from the Sun.

As for Earth's magnetosphere, on the side opposite the Sun the magnetotail extends far downstream of the solar wind, extending back more than 16 times the distance between Earth and the Moon. Compressed by the solar wind to within a quarter of the distance to the Moon on the side facing the Sun, it washes across the face of our nearest celestial companion when the Moon is at full and on the opposite side of Earth to the Sun, downstream of the solar wind. And this is another marvel of the geometry of our Earth–Moon system.

When the Moon enters Earth's magnetospheric tail close to full Moon, it is directly under the influence of the magnetosphere and the effects of this were measured by instruments left on the Moon at five of the six Apollo landing sites. Moreover, the electrical link between Earth and Moon has been measured by some scientists to change the potential of Earth's electrical field, a consequence of placing an iron-rich body in the tail of Earth's magnetic field. This, in turn, has been coincident with measurements made in the cerebellum and in neural paths of some animals and some humans.

But the solar wind also has a direct connection with Earth, to the joy and amazement of humans on its surface – the aurora borealis in the northern hemisphere and the aurora australis in the southern sky. Generally occurring within 10–20° of the

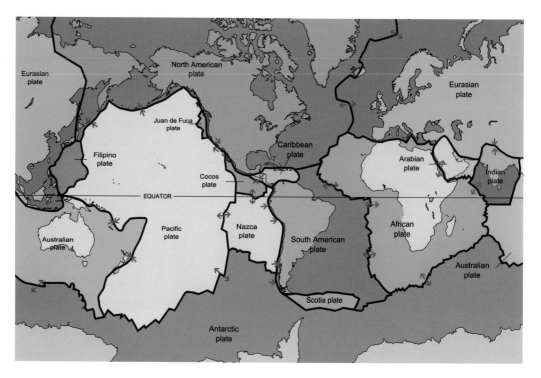

geomagnetic pole, and spanning a band of 1–2°, these remarkable light displays are caused when a level of quiet Sun activity injects solar particles directly into Earth's magnetic field lines, causing them to spiral down one of the points of polarity. It can also be caused by high and intense interactions with the planet's magnetic field lines, also sending particles down into the atmosphere, a reminder of just how close these otherwise harmful particles are to life on Earth.

When we see these aurora with our eyes we are watching emitted photons from ionised oxygen or nitrogen atoms in the upper atmosphere more than 50 miles (80km) above the surface, changing from an excited state to a ground state. Oxygen emissions are green or an orange-red hue according to the quantity of energy absorbed, while nitrogen emissions are blue and red depending on whether the atom is gaining an electron or returning to a ground state.

The very presence of the magnetosphere protects Earth and its atmosphere from the pressure of the solar wind which, over time, would leave it an airless planet were it not for the cocoon in which all living things are protected from the inevitable consequences of an unshielded and hostile environment. Radiation from the Sun would quickly strip away any primordial atmosphere which developed, or which was left over from the formation of Earth. The magnetosphere has provided an environment above the surface of the planet for the evolution of all living things, without which there would be no life outside deep oceans and rock niches.

The continents

One of the inclusive features on all the terrestrial planets in the solar system is that the crust is segmented into separate tectonic plates. On Earth, lighter materials separated out from the denser materials have tended to sink back into the mantle, a process which has been going on for four billion years. In this regard, as the only planet in the solar system known to have active tectonic plates, Earth is unique. This was not fully understood until the space age, when spacecraft were able to send back images of other worlds, land on their surfaces and observe features from orbit using remote sensing instruments to decipher their mineralogical codes. On no other planet is there irrefutable evidence of moving plates.

The notion of shifting continents itself is relatively new. Although the idea, then known as *continental drift*, had been around since the beginning of the 20th century, few scientists would give it credence, and very few were bold enough to support it. The logical fit of coastlines several thousands of miles apart introduced the intriguing possibility that they had somehow split and been carried apart to their present locations. How this could have been possible was completely unknown and a variety of theories abounded, including the general expansion of Earth from inside, tearing the crust as it expanded and rending it apart.

Then came evidence from the fossil records, where animals unique to certain continents appeared to have had a common origin but to have evolved separately in different places unconnected by land. It was as though they had originated when continents were connected, but gone along different evolutionary branches after they separated. Further evidence from geologists seemed to link certain features on separate continents, and a classic example of that was the Caledonian orogeny (mountain range) which is found in eastern North America and Scotland. We now know this is because they were once joined as a single feature before North America separated from the Eurasian plate when the Atlantic Ocean opened up.

By the 1960s and the dawn of the space age studies of Earth relied mostly on

ABOVE The US National Oceanic and Atmospheric Administration (NOAA) conducts undersea research using remotely operated vehicles such as Hercules, seen here prior to a descent. Robotic exploration of the deep was an unimagined capability a century ago. *(NOAA)*

BELOW Mudlogging is an essential part of the oil industry, from where a considerable amount of subsidised research goes on to better understand Earth's outer structure, feeding directly into a better scientific understanding of the planet. *(David Baker)*

examination and investigation of land and inner coastal waters. The technology did not exist for extensive surveys of the oceans and there was very little knowledge about seafloor features or understanding of how the floor of the ocean developed, let alone what it comprised. By the end of that decade detailed satellite imagery of Earth ran in parallel with an extensive exploration of the seas and oceans using sophisticated technology and sensors previously unavailable to scientists and oceanographers.

Because of the rapid development of submarine warfare, and the deepening political tensions between East and West, scientific investigation into the oceans was largely conducted by scientists working for the military. Ballistic missile-carrying submarines became one of the most closely guarded secrets of the Cold War and a rich harvest of data about the topography of the seafloor was classified for national security reasons. Submarines needed to know where they could hide, behind deep ocean ridges, trenches and undersea mountains to evade detection. A wealth of data mapped the ocean floor but almost all this information was classified and the scientific investigation of the oceans stalled until certain universities and institutions, primarily in the United States, began to fund their own research.

In the late 1960s the American National Science Foundation and the University of California Scripps Institution of Oceanography sponsored design and construction of a dedicated research vessel for deep-sea research. It was named *Glomar Challenger*, after Global Marine, the owner, and HMS *Challenger*, a British oceanographic survey ship of the 19th century. For 15 years beginning in August 1968 the *Glomar Challenger* conducted

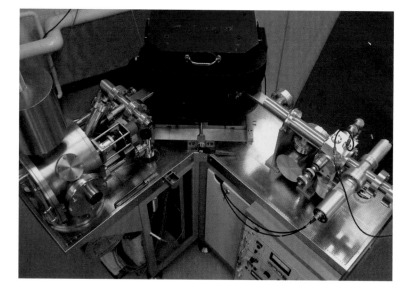

a comprehensive survey of the world's oceans, taking core samples using its deep drilling capabilities and providing scientists with a mass of information, a lot of which was also of use to the military in its quest for supremacy in the world of underwater warfare.

But a revolution had begun. An understanding of the way the floor of the seas and oceans moves is as much to do with an understanding of land tectonics as is the geological examination of the continents themselves. Not only did science gain valuable direct data from core samples extending deep into the seabed, a mechanism for studying when various undersea events took place was also discovered. By measuring the magnetic polarity of the core samples from precise locations it became possible to build up a map that revealed movement in the floor itself and to demonstrate through direct observation how the seabed had moved the continents around. This played directly into the hotly contested claims for continental drift, which quickly became the recognised science of plate tectonics.

Since the end of the Cold War in the early 1990s a considerable amount of information has emerged from previously classified files, and a number of scientists working for the military establishments of leading seafaring countries have added valuable information, as well as their experience and skill, to a sustained understanding of the world's oceans. While established now as a fact, the movement of the continents is also subject to direct measurement using satellite and space-based technology to measure the separation rate of various continental plates. Just as the gradual separation of Earth and the Moon can now be measured by laser retro-reflectors left on the lunar surface at three of the six landing sites, so too can similar technology be used to observe the much faster separation rate of continents on Earth.

Direct measurements have displayed some astonishingly high separation rates, the west Atlantic Ocean plate, for instance, moving at 3.4cm (1.3in) per year while the west Pacific Ocean plate is in some locations moving at 10.2cm (4in) a year. By contrast, the Indian subcontinent is burying itself into the area

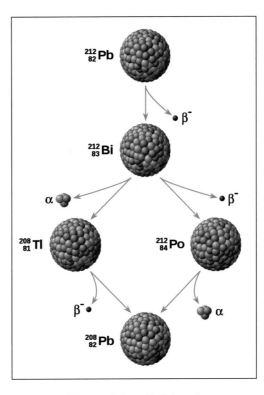

LEFT The radioactive decay chain from lead-212 where each nucleotide decays into a daughter product via a beta-decay where the final product is lead-208 (^{208}Pb), a stable isotope which cannot decay further. Measurements such as this are key to calculating elemental quantities in the past.
(David Baker)

underlying Tibet at 3.6cm (1.4in) each year. These are not insubstantial values, where intuitive logic creates a sense that geological forces move exceedingly slowly. Not so the crustal plates of Earth today. This begs the question as to when all this began and why, which returns us to the dawn of the Archean eon 4 billion years ago.

Summary

- Nuclear reactions occur naturally in sufficient concentrations and isotopic abundance.
- The crust is primarily formed from eight minerals separated out at differentiation after accretion.
- Minerals are rated according to definable key characteristics.
- Rocks are defined by category as igneous, metamorphic or sedimentary.
- Separation of ocean crust and continental crust was crucial to maintaining a dynamic planet.
- The Late Heavy Bombardment reset Earth's clock and rearranged the solar system.
- The magnetosphere is crucial for the development of life.
- Only in the 1960s could scientists verify the theory of plate tectonics.

Chapter Four

Geology in motion

The oldest rocks on Earth are found in the Acasta gneisses in Canada, age-dated to 4.031 billion years, but there are numerous examples of the ancient crust dating earlier than 3.6 billion years to be found on the North American continent, in Western Australia and in South Africa. Archean crustal material is exposed at the surface in areas where ancient terranes (dominant rock types) are distributed.

OPPOSITE Typical Banded Iron Formation (BIF) in Dales Gorge, Western Australia. BIFs are believed to have formed in oxygen-rich seas enriched with photosynthetic cyanobacteria. *(Graeme Churchard)*

INTERNATIONAL CHRONOSTRATIGRAPHIC CHART

www.stratigraphy.org   International Commission on Stratigraphy   v2015/01

Eonothem / Eon	Erathem / Era	System / Period	Series / Epoch	Stage / Age	GSSP	numerical age (Ma)
Phanerozoic	Cenozoic	Quaternary	Holocene			present
						0.0117
			Pleistocene	Upper		0.126
				Middle		0.781
				Calabrian		1.80
				Gelasian		2.58
		Neogene	Pliocene	Piacenzian		3.600
				Zanclean		5.333
			Miocene	Messinian		7.246
				Tortonian		11.63
				Serravallian		13.82
				Langhian		15.97
				Burdigalian		20.44
				Aquitanian		23.03
		Paleogene	Oligocene	Chattian		28.1
				Rupelian		33.9
			Eocene	Priabonian		37.8
				Bartonian		41.2
				Lutetian		47.8
				Ypresian		56.0
			Paleocene	Thanetian		59.2
				Selandian		61.6
				Danian		66.0
	Mesozoic	Cretaceous	Upper	Maastrichtian		72.1 ±0.2
				Campanian		83.6 ±0.2
				Santonian		86.3 ±0.5
				Coniacian		89.8 ±0.3
				Turonian		93.9
				Cenomanian		100.5
			Lower	Albian		~ 113.0
				Aptian		~ 125.0
				Barremian		~ 129.4
				Hauterivian		~ 132.9
				Valanginian		~ 139.8
				Berriasian		~ 145.0

Eonothem / Eon	Erathem / Era	System / Period	Series / Epoch	Stage / Age	GSSP	numerical age (Ma)
Phanerozoic	Mesozoic	Jurassic	Upper	Tithonian		~ 145.0
				Kimmeridgian		152.1 ±0.9
				Oxfordian		157.3 ±1.0
			Middle	Callovian		163.5 ±1.0
				Bathonian		166.1 ±1.2
				Bajocian		168.3 ±1.3
				Aalenian		170.3 ±1.4
			Lower	Toarcian		174.1 ±1.0
				Pliensbachian		182.7 ±0.7
				Sinemurian		190.8 ±1.0
				Hettangian		199.3 ±0.3
		Triassic	Upper	Rhaetian		201.3 ±0.2
				Norian		~ 208.5
				Carnian		~ 227
			Middle	Ladinian		~ 237
				Anisian		~ 242
			Lower	Olenekian		247.2
				Induan		251.2
	Paleozoic	Permian	Lopingian	Changhsingian		252.17 ±0.06
				Wuchiapingian		254.14 ±0.07
			Guadalupian	Capitanian		259.8 ±0.4
				Wordian		265.1 ±0.4
				Roadian		268.8 ±0.5
			Cisuralian	Kungurian		272.3 ±0.5
				Artinskian		283.5 ±0.6
				Sakmarian		290.1 ±0.26
				Asselian		295.0 ±0.18
						298.9 ±0.15
		Carboniferous	Pennsylvanian	Upper	Gzhelian		303.7 ±0.1
					Kasimovian		307.0 ±0.1
				Middle	Moscovian		315.2 ±0.2
				Lower	Bashkirian		323.2 ±0.4
			Mississippian	Upper	Serpukhovian		330.9 ±0.2
				Middle	Visean		346.7 ±0.4
				Lower	Tournaisian		358.9 ±0.4

THIS SPREAD The chronostratigraphic chart is the index to events in the history of Earth, its precise structure changing with time as more information became available. New data can also slightly change the dates attributed to specific parts of the chart, whether periods or ages, and therefore slightly older versions to that shown here may be out of date. This chart is current to January 2015 and has been produced by the International Commission on Stratigraphy. *(ICS)*

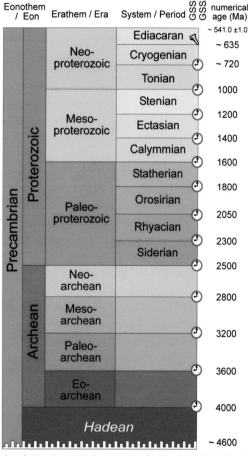

Units of all ranks are in the process of being defined by Global Boundary Stratotype Section and Points (GSSP) for their lower boundaries, including those of the Archean and Proterozoic, long defined by Global Standard Stratigraphic Ages (GSSA). Charts and detailed information on ratified GSSPs are available at the website http://www.stratigraphy.org. The URL to this chart is found below.

Numerical ages are subject to revision and do not define units in the Phanerozoic and the Ediacaran; only GSSPs do. For boundaries in the Phanerozoic without ratified GSSPs or without constrained numerical ages, an approximate numerical age (~) is provided.

Numerical ages for all systems except Lower Pleistocene, Permian, Triassic, Cretaceous and Precambrian are taken from 'A Geologic Time Scale 2012' by Gradstein et al. (2012); those for the Lower Pleistocene, Permian, Triassic and Cretaceous were provided by the relevant ICS subcommissions.

Coloring follows the Commission for the Geological Map of the World (http://www.ccgm.org)

Chart drafted by K.M. Cohen, S.C. Finney, P.L. Gibbard (c) International Commission on Stratigraphy, January 2015

To cite: Cohen, K.M., Finney, S.C., Gibbard, P.L. & Fan, J.-X. (2013; updated) The ICS International Chronostratigraphic Chart. Episodes 36: 199-204.

URL: http://www.stratigraphy.org/ICSchart/ChronostratChart2015-01.pdf

Large areas of present continental plates are known to have underlying rocks from the Archean eon and these are found across large swathes of eastern South America, Siberia and sub-Saharan Africa.

The Acasta gneisses are located across a 12.5 square mile (20km²) area in the Northwest Territories of Canada formed from magmatic rocks known as tonalite-trondhjemite-granodiorite, or TTG. These belong to the potassium-rich granitoid group that consists of silica-rich plutonic rocks but which differ in the abundance of plagioclase feldspar and alkali feldspar. There is debate about the origin of these materials, some scientists believing them to have a different origin to the classic one of subduction magmas, where ocean crust dips back below lighter continental crust. Others believe they were formed in the lithosphere, the outer mantle, by mantle plumes, upwelling of thermal hotspots from deep in the mantle itself.

Whatever their precise origin these Archean terranes are sometimes referred to as shields or cratons, one of the oldest being that in Canada's Northwest Territories. Less than 600 million years after the accretion of Earth, the Sun was only 75% as potent an energy source as it is today. Were it not for a dense atmosphere raising the temperature, the entire planet would wait a further billion years before the surface was above the freezing point of water. But Earth itself produced almost four times as much heat as it does today, and even by the end of the Archean eon heat flow was still twice the present rate.

Thermal energy came from three sources: planetary accretion and the residual heat left over from that; latent heat generated by crystallisation in Earth's core; and radioactive decay primarily from uranium (^{235}U and ^{238}U), potassium (^{40}K) and thorium (^{232}Th). Some radiation too came from radiogenic elements with short half-lives and which were decaying out. Over time Earth has cooled down, energy from the accretion process having now been absorbed and many of the radioactive elements having decayed too. Today Earth produces 42TW (one terawatt equals 1,000 billion watts), of which three-quarters comes from radioactive decay.

Back at the beginning of the Archean, the upper mantle was at least 212°F (100°C) hotter than it is at present and this would have resulted in a different type of crustal material to that with which we are familiar today. Key to understanding this early phase are several unique rock types that hold clues to how the early crust would have behaved. One of those is an ultramafic volcanic rock derived from the mantle known as komatiite. Mafic rocks are silicate in base composition and rich in magnesium and iron, deriving their name from 'magnesium' and 'ferric'. Ultramafic rocks have less than 45% silicon and above 18% iron with

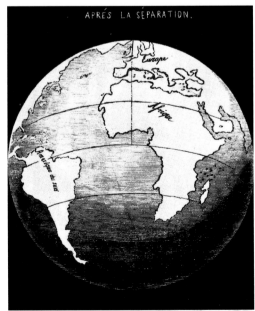

mafic materials constituting more than 90% of their composition.

Komatiites are very rare and can only be found in Archean rock shields and greenstone belts, the latter being zones where ultramafic rocks are associated with ancient cratons and are rich in chlorite, actinolite and green minerals. Komatiites are found in great quantities with ages greater than 2.5 billion years but nothing since then. They were generated when magma crystallised during partial melting below a depth of 75 miles (120km) and with eruption temperatures of 2,777°–3,000°F (1,525–1,650°C), much hotter than ocean basalts today which erupt through mid-ocean ridges at 2,280°–2,460°F (1,250°–1,350°C).

The temperatures in the mantle before 2.5 billion years ago would have resulted in 50–60% of it melting, which would have been the reason why the komatiites are so abundant. Today, no more than 30% of the mantle is molten which is why the ocean crust is basaltic. Basalts are igneous rocks containing less than 20% quartz and are less than 10% feldspathic by volume, typically containing interspersed minerals and with an average mean density of $3g/cm^3$. Scientists know nothing about the precise mechanism that produced these ultramafic komatiites but

the most logical explanation is that they were produced in very hot plumes, about which more later. But that is not the point, since they could only have been formed under these extremely hot conditions, as laboratory experiments and calculations reveal, thus pegging a band of temperatures that the Archean era experienced. Time to return to those tonalite-trondhjemite-granodiorite rocks.

In similar fashion to the komatiites, the existence of the Archean TTGs takes geologists back to a time now past. They have well-defined mineralogical and chemical composition and, being granitoids, they consist of quartz, plagioclase feldspar, a form of black mica called biotite and amphibole. In this they are very different from today's continental crust, which has granodioritic to granitic composition, with quartz, alkali feldspar, plagioclase feldspar and biotite. In this difference, the TTG rocks are sodium-rich compared to the potassium-rich continental crust of today. One other vital difference is that the continental crust is rich in heavy rare earth elements while the TTGs have low abundances of these elements.

What this shows, and what has been verified through laboratory experiment, is that the original Archean crust was very different to the continental crust that appeared

ABOVE Cratons in West Gondwana, a unified supercontinent which would emerge from aggressive plate tectonic mobility across the asthenosphere.
(David Baker)

ABOVE LEFT Komatiite lava in rocks from South Africa which have helped unlock the activities of the early Earth and the chemical composition of the early crust, which would change in the Archean eon.
(CSIRO)

later. All analyses show that TTGs can only be reproduced in the presence of residual garnet and hornblende and that this must be accomplished at depths below 25 miles (40km) where the pressure is 12,000 bar. The material from which the TTGs originate must have been generated by partial melting of hydrated basalts, and the circumstances inferred from the komatiites confirm the general theory that the early crust was very different.

To understand how the early crust differs from the later crust, and to unravel the forensic analysis of how the present crust originated, it is helpful to briefly review what we know for certain today about the way the crustal plate system works.

With a mean density of $2.7g/cm^3$, continental crust is lighter than ocean basalt crust, which has a mean density of $2.9g/cm^3$, and rides on top of the lithosphere, which has a density of $3.3g/cm^3$. The lower mean density means that the crust rides across the mantle pushed along by slightly denser ocean basalts which extrude from depth at the mid-ocean ridges and separate in opposing directions creating a divergent zone, pushing the lighter continental material along. At the boundary with a continental plate, the denser ocean basalts sink back down into the mantle at subduction zones where the material returns to the mantle, eventually to be recycled back up at the mid-ocean ridges to start the process over again.

This conveyor-belt motion allows the continents to perpetually move around the planet, and as we shall see in Chapter 7, this factor alone could have been responsible for the development of advanced life. However, the ocean basalts take no more than 60–180 million years to complete their cycle from ridge to subduction. At the mid-ocean ridge system the basalts are produced through partial melting

of an anhydrous mantle, and because they are themselves also anhydrous (with less than 0.3% water) this causes crystallisation of minerals such as olivine (Mg_2SiO_4), orthopyroxene ($Mg_2Si_2O_2$), clinopyroxene ($CaMgSi_2O_6$) and plagioclase ($CaAl_2Si_2O_8$). The mid-ocean ridge is also heavily faulted and this draws water down into the cracks to hydrate the minerals. These are known as serpentine antigorite, chlorite, talc and hornblende amphibole, where the water content has increased to 1.5–7%.

Water is vital, for the temperature at which a basalt will start to melt depends on the amount of water present. At a pressure of 10,000 bar, typical at a depth of 19 miles (30km), the anhydrous basalt will start to melt at a temperature of 2,190°F (1,200°C). In a hydrated state with the basalt in the presence of 5% water, that melt temperature is 1,112°F (600°C). The greater the quantity of water, the lower the melt temperature. The presence of extruding basalt in an aqueous environment not only hastens and eases the melt process, the water in such large quantities as those at the ocean floor quenches the temperature and cools the rock.

When the cold basalt of ocean crust encounters the subduction zone at the continental plate boundary, it only gradually acquires higher temperatures. The geothermal gradient at the boundary between the subducted ocean crust and the mantle, known as the Benioff plane, is relatively high. The process of altering the ocean basalt as it sinks further into the mantle follows a comparatively benign thermal gradient, quite the reverse of when it started its journey at the mid-ocean ridge. At a depth of about 50 miles (80km), this gradient encounters the temperature/pressure line where antigorite forms, then chlorite, talc and finally hornblende. At this point the minerals are no longer stable and metamorphose into new minerals that will be stable at these new temperature and pressure regimes.

The sum product of this chemical reversal is to return the rock to an anhydrous state from which it originated and at a depth of 75 miles (120km), and a temperature of 750°F (1,380°C), the rock reaches a completely solid state and is totally anhydrous. At this state, the anhydrous melting point of a total solid is 2,550°F (1,400°C), but the rock never sees this

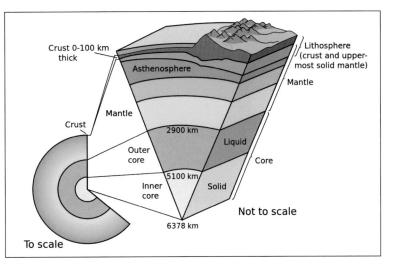

ABOVE A graphic depiction of a pie-slice of Earth from the crust to the core, exhibiting the different core sections and inner and outer mantle. It should not be inferred from this that separation planes between the various divisions of mantle and asthenosphere are at a consistent distance from the centre of Earth. Irregularities and inhomogeneity playing a big art in determining actual depth in various places. (David Baker)

BELOW Isostasy is the determining factor in the ability of continents to grow large and support major structures such as mountains. Here, the elevated structure of the mountain anti-root (h_1) is brought to isostatic buoyancy by the root (b_1). The "Moho" is the Mohorovocic discontinuity between the upper mantle and the ocean crust, a place where seismic waves increase in velocity with depth. (David Baker)

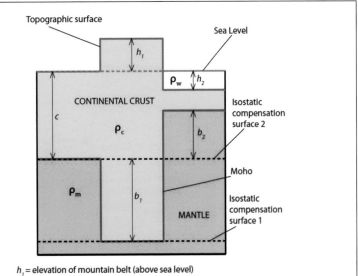

h_1 = elevation of mountain belt (above sea level)
h_2 = depth of marine basin (below sea level)
b_1 = thickness of crustal roots (below depth of Moho in a cratonic area)
b_2 = thickness of lithosphere mantle bulge (above depth of Moho in a cratonic area)
c = thickness of continental crust in an undeformed (cratonic) area (ca. 35 km)

ρ_w = density of sea water (ca. 1,000 Kg/m^3)
ρ_c = density of continental crust (ca. 2,800 Kg/m^3)
ρ_m = density of mantle (ca. 3,300 Kg/m^3)

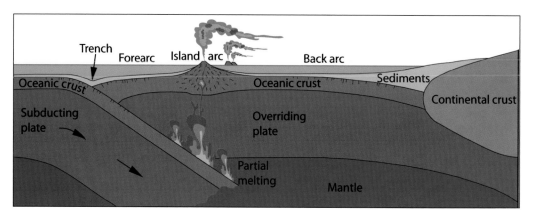

temperature level and so it returns back into the mantle as a dry and solid slab. As this dehydration process is under way on the slowly descending basalt, the water rises up through the lithosphere to the mantle wedge, that part of the mantle that lies across the underside of the continental

subduction zone. This increases the water content of that zone, melting it due to the reduced hydrous melt temperature where it serves as a lubricant for oceanic crust returning toward the mantle. Now we can return to where it all began.

Reminding ourselves that the magma that produced the TTGs was in a stable equilibrium with a residue containing garnet and hornblende, it becomes clear that the Archean crust was very different, while being produced in a similar manner to the later continental crust on which we all live today. It was precisely because Earth was hotter and the recycling episodes more frequent and occurring at a faster rate than today, that there was difficulty identifying the conditions that produced the present continental crust. What is certain from the record in the rocks and minerals is that the mantle was 212–390°F (100–200°C) hotter

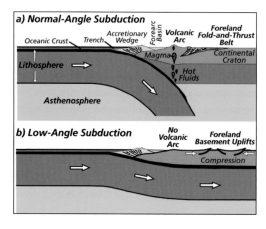

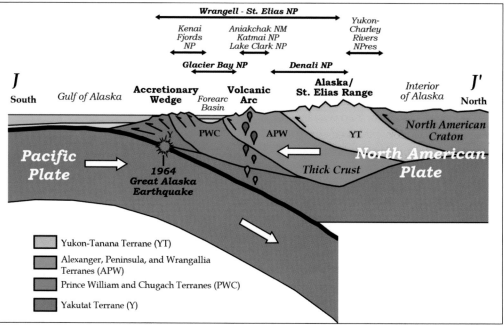

than it is today, which would have reduced its viscosity by a factor of 10.

Viscosity is a measure of the resistance to deformation through stress: the higher it is the more resistant it is to that stress, either shear or tensile. A lower viscosity in the mantle would have made it more slippery and it would have been less resistant to the structural mass of the crust above it. Evidence for the existence of very large cratons from this early period is irrefutable, and this leaves open the issue as to how they could have remained 'floating' on the mantle when it was much less viscous than today. One way of accommodating the large features on the Archean crust is to believe that the crust was hotter and more ductile – its ability to deform under stress – and therefore less likely to thicken up and remain intact.

We have already seen that every mineral has a very particular range of temperatures and pressures in which they remains stable, and that if these conditions change new minerals form and crystallise in a manner that does not change the overall chemical composition of the rock but redistributes the elements into new phases. In this way it is possible to tell to which particular process of metamorphism the rock has been exposed. Detailed analysis and measurement of rocks from many early cratons shows that the temperature gradients in the crust were the same as they are today, which may appear surprising since we have already demonstrated that the overall heat output from Earth was much greater than it is now.

The reason for this seeming paradox is quite simple. While the overall thermal energy budget was higher, the thermal load on specific areas was between wide extremes. The process of convection that drives thermal energy from mantle convection cells to replenish ocean crust at the surface (on the seabed) is the most powerful form of heat transfer possible. Both conduction and radiation – the only other two ways of moving heat around – are much less efficient. Convection drives the mid-ocean ridges but conduction is responsible for moving the heat everywhere else. This is why the rate of convective upwelling from the mantle can be much greater while the static Archean crust was influenced by less efficient physical forces exposed on the surface and therefore balanced the overall heat load.

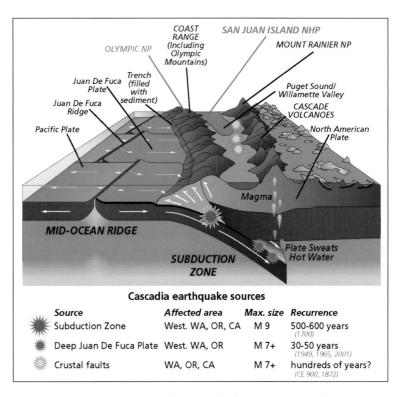

Cascadia earthquake sources			
Source	Affected area	Max. size	Recurrence
Subduction Zone	West. WA, OR, CA	M 9	500-600 years (1700)
Deep Juan De Fuca Plate	West. WA, OR	M 7+	30-50 years (1949, 1965, 2001)
Crustal faults	WA, OR, CA	M 7+	hundreds of years? (CE 900, 1872)

ABOVE In another depiction of plate subduction, transform faulting can be seen along the offshore region parallel to the Olympic Mountains with associated volcanic activity and crustal faulting. Earthquake probability is also shown. *(David Baker)*

BELOW Anticlines (A) are structural deformation of previously undeformed layers where older uplifted material is flanked by younger synclines (B). Complex interactions of both convey veined waveforms (C) while intrusions cause plunging folds (D). *(David Baker)*

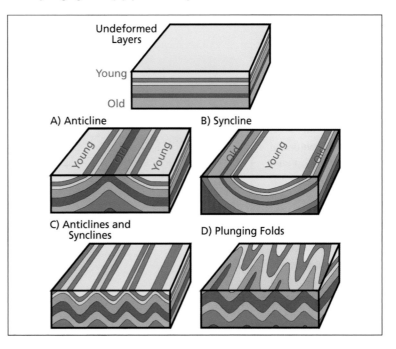

Evidence from the early cratons, particularly those in South Africa, show that the thickness of the crust was about 25 miles (40km), which is about what its successor is today. Their rocks exhibit garnets formed under pressures of 15,000 bar, which compute to that depth. Today the average depth of the continental crust is 19 miles (30km), at times extending to a depth in excess of 43 miles (70km) where great mountain ranges are present and the average elevation is a mere 1,000ft (300m). Because of isostasy, defined as a state of equilibrium between the crust and the mantle, a vertical elevation above the mean surface is known as the anti-root and the stabilising area of equilibrium below the mean surface level is known as the root.

Because the mass above must be stabilised through a greater distribution of surface area across the root to conform to this physical law, isostatic equilibrium requires in general that mountains above ground must have a greater volume below. This is the geologist's equivalent of the floating iceberg, but for a different physical principle conforming to the same law of isostasy. Not all mountains have an anti-root, the Himalaya being one such example, where the latent energy of the Indian plate moving up and into the Asian plate has created a state of disequilibrium that will only dissipate when there is no relative motion between the two. Another example is the Alps, where Africa is moving up into continental Europe.

Collisions between continental masses form mountains – there is no other cause – and can be thought of as seam-welds between two lighter plates, brought together by horizontal forces driven by the motion of ocean basalt far away, buckling the crust in upthrusts as the two are brought into contact. One example is the Ural Mountains, a 1,600-mile (2,600km) range running north to south from the coast of the Arctic Ocean to north-western Kazakhstan, used as a geographical marker between Europe and Asia. The Urals were formed about 280 million years ago, very recent in geologic time, during a period of great crustal readjustment on Earth which we will come to later.

The big question remains. Just how much of the Archean crust remained as the continental crust of today formed? There is strong evidence from Isua in Greenland that the continent there is at least 3.87 billion years old. But it is here that the controversy over the formation of the continental crust rages most fiercely, advocates of two opposing theories fighting it out over the same geologic formations and the same data.

One theory has it that the Archean planet was too hot internally, and the viscosity too low, the material flowing too freely for the mantle to have generated a crust that could form isostatically stable slabs of lighter materials.

RIGHT The ocean crust is measured by age in years from eruption at mid-ocean ridges, the oldest being at the continental margins where subductions zones are found. Note the aged eastern Mediterranean sea floor. *(Elliot Lim)*

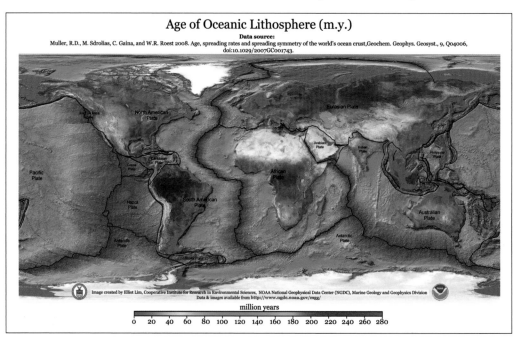

Age of Oceanic Lithosphere (m.y.)
Data source:
Muller, R.D., M. Sdrolias, C. Gaina, and W.R. Roest 2008. Age, spreading rates and spreading symmetry of the world's ocean crust,Geochem. Geophys. Geosyst., 9, Q04006, doi:10.1029/2007GC001743.

Image created by Elliot Lim, Cooperative Institute for Research in Environmental Sciences, NOAA National Geophysical Data Center (NGDC), Marine Geology and Geophysics Division
Data & images available from http://www.ngdc.noaa.gov/mgg/

million years
0 20 40 60 80 100 120 140 160 180 200 220 240 260 280

This theory has it that the crust was simply too weak and pliable to support mountains and high elevations. Its proponents say that the mantle/crust interaction was dominated by vertical convective cells pushing up into the crust and replenishing it, citing the komatiites as supporting that because this would have favoured their formation. Moreover, they say, the TTGs would have been produced through partial melting of basalts that became detached from the underside of thick ocean crust, not unlike today's ocean plateaus.

The alternative theory suggests that because the processes were different the anomalies can be accommodated within the available evidence: that horizontal movement of plates was already taking place, as evidenced by the rock strata at Isua where the oldest terranes exhibit unequivocal evidence for that. The truth may lie somewhere between the two, allowing for horizontal continental displacement as Earth displays today, but not discounting the early multi-plume model of the alternate theory, that numerous convective cells were punching up vertically without lateral or horizontal motion. And while there are these two theories about the early crust, there is another way to explain the process bridging Archean to Proterozoic crust.

To find that way it is once again necessary to go back and refer to the TTGs and komatiites that display anomalies in their production compared with geologic possibilities today. Because the thermal output of Earth's heat engine was four times its present level, some explanation has to be found for why that heat was not retained in the planet but released, possibly through many, rather than a moderate number of, mid-ocean ridges. In some places on Earth the heat flux today is at a level equivalent to the Archean average for the entire planet: 240mW.m^{-2}. One such place is the North Fiji basin where the number of active ridges is very high, possessing a linear length 20 times the collective linear length of all the other ocean ridges across the Pacific.

The result of this is that the North Fiji basin has many micro-plates, an abnormally high number that drives up the overall thermal load per given area to a level that appears to replicate the sort of intensity that could have

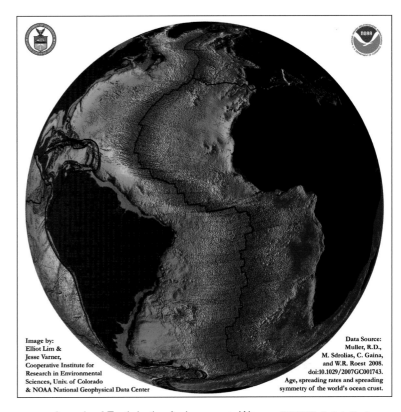

Image by:
Elliot Lim &
Jesse Varner,
Cooperative Institute for
Research in Environmental
Sciences, Univ. of Colorado
& NOAA National Geophysical Data Center

Data Source:
Muller, R.D.,
M. Sdrolias, C. Gaina,
and W.R. Roest 2008.
doi:10.1029/2007GC001743.
Age, spreading rates and spreading
symmetry of the world's ocean crust.

covered much of Earth in the Archean eon. We have already seen that the high geothermal gradient along the Benioff line encountered by hot ocean crust was relatively high. Named after Hugo Benioff, the scientist who worked out a way of determining whether a sequence of earthquakes was produced by a single fault or several faults, it states that the square root

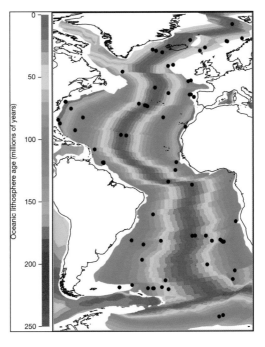

ABOVE A detailed view of the North and South Atlantic sea floor with multiple transected faults where pivotal points of rotation for North and South America cause torsional stress. *(Elliot Lim)*

LEFT Another view of the ageing Atlantic Ocean sea floor with coastal detail and black dots indicating where samples have provided accurate dating. *(NOAA)*

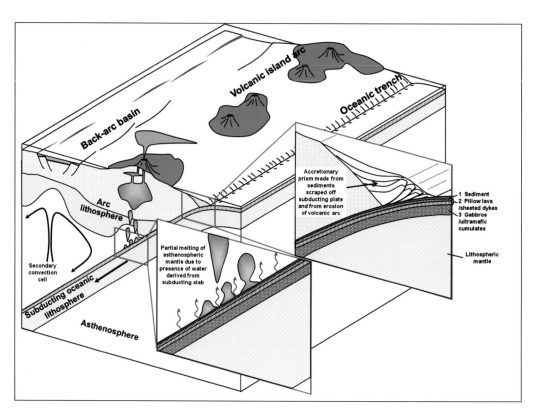

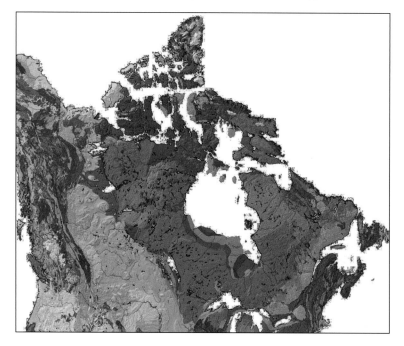

of an earthquake's energy is proportional to both the elastic rebound strain and the rebound displacement.

The Benioff seismic zone is determined by the angle at which the subducting slab of crust, which is a product of the negative buoyancy of the slab and forces flowing from the asthenosphere, encounters the upper mantle.

With a younger and more buoyant hot lithosphere the angle is shallow, whereas a colder, denser and older lithospheric slab will cause steeper dips ranging from 30–60°. Because it is determined entirely by the constraints of subduction angle, temperature and density, the Benioff plane can exist from close to the surface down to a depth of 416 miles (670km).

Subducting a young, hot lithospheric slab would provide the high thermal gradients at the Benioff plane and allow the hydrous melting of young basalts that provided the environment for creating the TTG magmas. But the angle of subduction would be very much shallower than the subduction planes today. This can be demonstrated in Ecuador at the juxtaposition of the Carnegie Ridge, which is being carried along by the older and cooler Nazca Plate, and the South-American Plate, where the Carnegie Ridge is subducting at an angle of only 20°. Because of the subduction angle the volcanic arc thus created is much larger, at 93 miles (150km), than is generally the rule today, where such arcs are typically 31 miles (50km). It has all the appearance of an Archean crustal geometry and may be one of the few places on Earth where the world of 2.5–4 billion years ago can still be seen and measured.

The two theories attempting to describe the formation of the continental crust from the Archean crust can be united by the concept of sagduction, a form of subduction but one in which tectonic activity is characterised by vertical movements. Driven by gravity it involves the sinking of surficial greenstone materials into narrow belts with the exhumation of deeper granitic crust into broad domes. Sediments and basalts bearing garnet collected from the rim of the East Pilbara Dome craton in Western Australia display higher pressure but lower temperature for equilibrium of 9,000–11,000 bar and 840–1,022°F (450–550°C). Samples collected from the core of the dome imply a formative 6,000–7,000 bar and 1,200–1,240°F (650–670°C). Measurements such as these that show two cycles of metamorphic activity suggest a more rapid flow of material. These samples are dated at 3.31–3.44 billion years.

Based on the evidence to date, it appears that as Earth has cooled gradually the nature and composition of the crust has changed, with characteristic differences in the subduction angle, which has become steeper as the cooling continued. Where once there was a global mosaic with numerous mini-plates and much greater length of mid-ocean ridge divergent zones

to release greater quantities of energy from the mantle, there now exist 17 major rigid tectonic plates and many smaller plates. The Archean world was replete with a large amount of volcanic activity as material was recycled frequently and at much shallower angles back into the mantle, from where it would be returned to the ridges.

Over the 1.5 billion years of the Archean, the

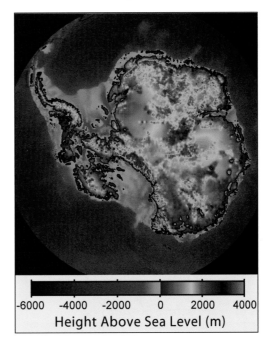

Height Above Sea Level (m)

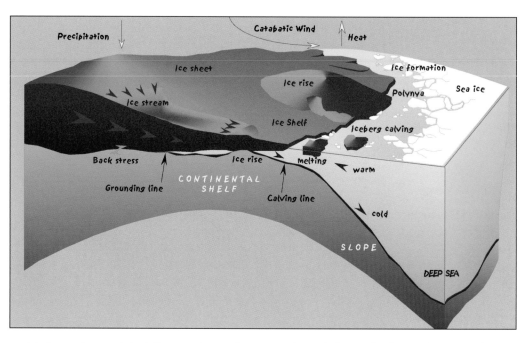

world changed a great deal. Progressive cooling brought an end to the komatiites by the start of the Proterozoic eon 2.5 billion years ago, the TTGs produced as a result of melting hydrated basalts giving way to granodiorites as the mantle-wedge melted and vertical sagduction came to an end. The lithospheric plates began to increase in size and the hydrothermal activity began to decrease, albeit very slowly. This was still a world unrecognisable today, the oceans broken by continental plates steaming slowly at a temperature of 158°F (70°C), a geologically active world in which volcanic eruptions were commonplace on and at the edges of the

continental plates, slowly growing in number and accumulating in plateaus.

As the subducted ocean crust melted at increasingly greater depth geothermal gradients decreased and by around 3.4 billion years ago the transition toward the Proterozoic, where the modern continental forms would be almost complete, was well advanced. The chemical changes which occurred to the ocean magmas was a logical sequence of the progressive changes to the planet as a whole, as the Sun became less intense and greater screening of ultraviolet was provided by a very dense atmosphere, resulting in a degree of protection from solar and cosmic radiation from the established magnetic field and the radiation belts.

Life was already starting to make its presence felt, but that is a story for Chapter 7; suffice it to say here that the high-temperature volcanic emissions, the komatiites, universal hydrothermal activity and great swathes of chemical sedimentation incorporating Banded Iron Formations (BIFs) eliminate any possible trace of living organisms. But they leave their evidence, as we shall see. It is not by inference alone that scientists can trace the origins of life here on Earth, and in so doing identify the markers that others may look for elsewhere, on other planets in the solar system. But as the story moves to the Proterozoic eon Earth encounters change on a colossal scale, a story found in the record of the rocks.

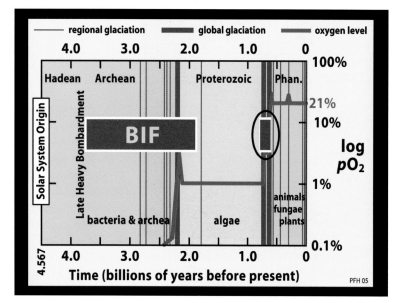

Transforming Earth

This section of the Earth story concerns the Proterozoic and occupies the longest column in the chronology of the planet, extending from 2.5 billion years to 541 million years ago, bridging the Archean world with that of the Phanerozoic, the last of the four great eons of time. For purposes of classification, this eon is further subdivided into three geologic eras: the Paleoproterozoic (2.5–1.6 billion years); the Mesoproterozoic (1.6–0.90 billion years); and the Neoproterozoic (0.90– 0.54 billion years).

This eon marks the transition in geological terms from a time of hot and less viscous surface magma pushing the continents around with instability and upheaval, to a sustained period of calmer activity, with continent building and more stable conditions.

Gone were the massive basaltic eruptions such as that marked by the Canadian Shield formed between 3.5 and 2.5 billion years ago. It has more than 150 volcanic vents and originally supported mountains 39,000ft (12,000m) high, now depressed through weathering, glaciation and subsidence to a relatively flat terrain. The Canadian Shield covers an area of 3.09 million square miles (8 million square kilometres) extending from most of Greenland to New York and down into Michigan and across to north-western Canada. Although this example is unique in that it is the largest identified shield, it is typical of the activity that characterised the later Archean, a period of gigantism which was to portend a less chaotic but more evolving sequence.

Stretching for just over two billion years in total, the Proterozoic covers the period of great change across the planet, massive global glaciations, oxygenation of the atmosphere, the first complex multi-celled life forms, green algae in the world's oceans, and great geophysical events bringing catastrophe to land masses while constructing continents. It includes the formation of the great Antarctic orogeny, the reduction of new minerals and the transformation of Earth into the world we know today, producing the first fossilised life forms.

Although the changes that were to characterise the Proterozoic configured the planet for the luxuriant distribution of flora and fauna that would define the Phanerozoic, the most dramatic activities associated with Earth's engine are defined by the building of large continents and supercontinents. Scientists are uncertain about the precise configuration of these continental masses during the two billion years after the Archean, with two opposing views: that there were several supercontinents that came together several times; or that there was one single continent until about 600 million years ago when it began to break up.

About one-third of the total surface area of Earth is formed of light continental crust, the rest is ocean crust formed from convective cells of magma rising up through mid-ocean vents, flanked by steep ridges, generally higher than the abyssal plains covering the vast expense of ocean floor between the ridges and the subduction zone at the edge of an associated continental mass. But whether this configuration of the 17 primary continental plates around now was a matrix of many more or just one for most of the Proterozoic is uncertain.

From the geological evidence, the mineralogical traces and the relative abundance of elements within the mantle and the crust it seems logical to conclude that the transition from the sagduction world at the end of the Archean to the present assembly of plates should see a gradual assembly of many plates into the few around now. A seminal contribution to the debate was made, some say settled, by the research work of John Tuzo Wilson.

LEFT John Tuzo Wilson, who added great authority to the debate on plate tectonics and seafloor spreading, placing the concept on a deeply scientific footing and uniting argument around the principles for its activity.
(John Wilson)

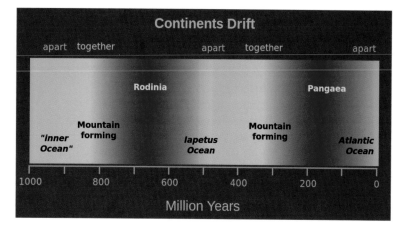

ABOVE The Wilson cycle helped put interdisciplinary science into the argument for plate tectonics, allowing forensic analysis of the geological record to demonstrate the repeated accretion of continental masses in periodic supercontinents. *(USGS)*

RIGHT Wilson developed the theory of convective cells pushing up through the asthenosphere and lifting overlying land masses, those now known as diapirs. *(David Baker)*

RIGHT Partial melting of the asthenosphere contributes to diapirs and also to elevated plume impingement as seen here where temperature is set against depth for a sequence of geological conditions and events. *(USGS)*

Born in 1908 in Ontario, Canada, 'Jock' Wilson was of Scottish and French descent, an avid traveller, photographer and geographical researcher. From studying geology at university in the mid-1930s, Wilson served in the Canadian Army during the Second World War before returning to his passion – the study of Earth and its dynamic forces. With the revolution in understanding that came from the new theory of continental drift, Wilson threw himself with passion into the study of plate tectonics, touring geologically interesting sites, raising theories and putting forward radical ideas that began to take hold.

Wilson was one of the first to accept plate tectonics head on. Utterly convinced of its veracity, he postulated theories which are today enshrined as accepted logic, maintaining that the Hawaiian Islands are a chain of volcanic domes formed as the plate is moving across a thermal plume in the mantle, pushing up lava every so often and creating a series of mini-volcanoes punching through the lithosphere. He was correct in that theory. He also gave rise to the idea of the transform fault, envisaging two adjacent and parallel plates moving past each other in a sliding motion. But perhaps his greatest contribution was to formulate an explanation for what is really happening with continents and their evolution.

Known as the Wilson Cycle, the sequence has

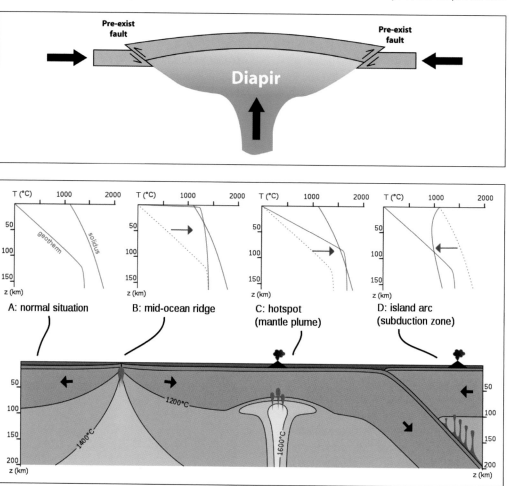

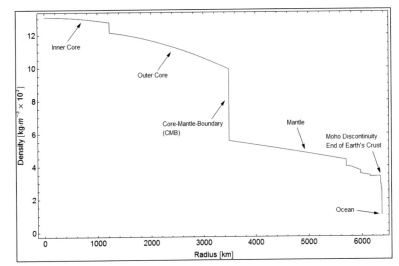

been accepted now as fact and helps place in context the various observations of this transition phase that characterised the Proterozoic. It is based on sound measurement and paleogeographic analysis of sea levels at various times. When continents were gathered together as a single platform sea levels were lowest, and when they dispersed the sea levels rose. The age of the oceanic lithosphere is a useful regulator on the depth of the ocean basins and in turn on global sea levels.

As the divergent magma emerges from the mid-ocean ridge and cools and shrinks, its density increases and its thickness decreases. This results in a lowering of the sea floor proportional to increasing distance from the mid-ocean ridge. To calculate what that effect will be, by taking a section of the ocean lithosphere less than about 75 million years old, and applying a simple model of conductive cooling, we can demonstrate that, where d is the depth of the ocean basalt and t is the age of that basalt,

$$d(t) = (2/\sqrt{\pi})\alpha_{eff}T_1\sqrt{kt} + d_r$$

where k represents the thermal diffusivity of the mantle lithosphere, α_{eff} represents the thermal expansion coefficient for rock, T_1 is the temperature of the ascending magma in comparison with the temperature of the upper boundary (1,120–1,220°C/2,048–2,228°F) and d_r is the depth of the ridge below the surface of the ocean.

The thermal diffusivity of the mantle lithosphere is approximately 8×10^{-7} m^2/s while the thermal expansion coefficient of rock is about 5.7×10^{-5} C$^{\circ-1}$. This can be simplified by loading in the sea floor values so that the equation can be

$$d(t) = 390\sqrt{t} + 2500$$

for the Atlantic Ocean and the Indian Ocean, where d is in metres and t is in millions of

years. Because sea levels rise when continents separate, or rift, stretching them in surface area, compressing the continents has the reverse effect, lowering sea levels and exposing continental shelves. Because these shelves are shallow and have low inclines, a small increase in sea level will have a major effect on the amount of land inundated by sea and ocean flooding. Conversely, a young ocean floor will create a shallow sea and if the ocean is old the level will be lower.

This was the fundamental equation that Wilson used to demonstrate that there is a relationship between the supercontinent cycle and the mean age of the sea floor. Running these equations produces some interesting

BELOW As mid-ocean ridges release basaltic magma, it cools down freezing in the magnetic polarity of the Earth, which switches poles at irregular intervals. By taking core samples and determining the polarity of the rock as extracted at a measured distance from the ridge, its age can be calculated. *(Chimee2)*

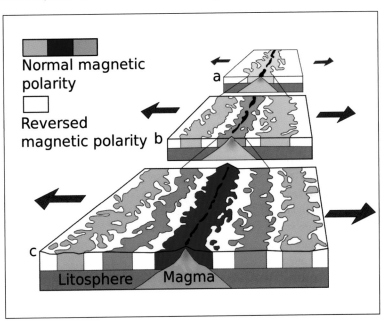

Normal magnetic polarity

Reversed magnetic polarity

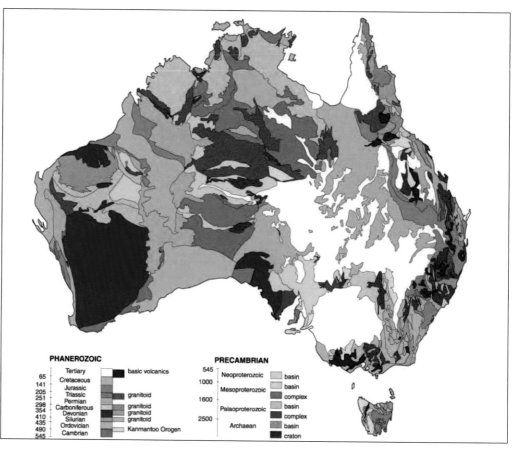

PHANEROZOIC		PRECAMBRIAN	
65 — Tertiary	basic volcanics	545 — Neoproterozoic	basin
141 — Cretaceous		1000 —	basin
205 — Jurassic		Mesoproterozoic	complex
251 — Triassic	granitoid	1600 —	basin
298 — Permian		Palaeoproterozoic	complex
354 — Carboniferous	granitoid		basin
410 — Devonian	granitoid	2500 —	
435 — Silurian	granitoid	Archaean	basin
490 — Ordovician			craton
545 — Cambrian	Kanmantoo Orogen		

consequences that we will explore further in Chapter 6 when we look at the evolution of the atmosphere, but insofar as they are impacted by Earth's engine, they are relevant here too. Because a single supercontinent implies an abundance of aged sea floor and low sea level, this will amplify the impact on climate because the continent itself will be the driving influence. This could force glaciation, which will further lower the level of the sea. Dispersed continents shift the controlling influence from the continental land mass back to the seas, where glaciation is unlikely and sea level is not lowered.

Before looking at a typical Wilsonian cycle, we return to the history of supercontinents and their presence on the globe. Combining both Wilsonian logic, mathematical equations and the fact that the continental material moves on a spherical Earth, it is axiomatic that if they disperse as separate plates on one side of the planet they will tend to converge on the other side. Of course, it is not that simple. Complex motions and divergent separation angles provide an almost unpredictable pattern of continental distribution and a lot of moving

around could ensue prior to reconnection. Conversely, a few simple movements away from the supercontinent status could bring separate continental plates together at different times and different rates. In fact that is the situation on Earth today.

On average, supercontinents tend to form every 400–450 million years and there are probability equations to calculate that frequency, laws driven by geological observation of pieces of old continental material around today. Just as the Archean structures could be differentiated, so too does the geologic record support some fairly accurate chronology as to very early supercontinent formation.

The earlier proposed supercontinent is known as the Vaalbara event, evidence coming from South Africa's Kaapvaal craton and the Pilbara craton in Australia, the latter being cited earlier when discussing sagduction and the formation of the continental crust from the thicker Archean crust. The name itself derives from the last few letters of each craton. Paleomagnetic evidence appears to show a juxtaposition between the two which,

although far apart today, were together more than 3 billion years ago. The earliest evidence in the cratons for clustering appears as early as 3.6 billion years ago, but the separation was complete 800 million years later. There is evidence for another supercontinent forming around 3 billion years ago known as Ur, others being Kenorland (2.7–2.1 billion years ago) and Columbia (1.8–1.5 billion years ago).

The first reasonably substantiated supercontinent bears the Russian name for Motherland – **Родина**, or Rodinia. It was first identified as such in 1990 by Mark and Dianna McMenamin, notorious for having espoused heavily controversial theories about a wide range of supposedly supernatural events and connections between racial and ethnic groups separated by great distances. But the work on Rodinia is based on geological evidence and analytical reconstruction, and while it is not proven to have existed in the interpreted form it is gaining acceptance as a logical precursor to the most recent supercontinent – Pangea.

Rodinia is believed to have begun forming as a single supercontinent of all Earth's land

masses around 1.1 billion years ago and to have lasted for 350 million years before breaking up 750 million years ago. It is thought to have formed from the accretion of fragments of previous supercontinents and to have constituted a single mass centred at and south of the equator. Postulating that Rodinia did not start breaking up everywhere simultaneously, some geologic evidence exists to show that rifting began widely, with large-scale volcanism occurring across most of what are today separate continents.

Lava flows and volcanic activity appeared to have forced the Rodinia supercontinent apart into two major land masses, rifting what is now Australia, eastern Antarctica, India, the Congo and the Kalahari on one side and the Amazonis region, West Africa and the Rio cratons on the other. Known as the Pan-African orogeny, this arrangement appeared to be in balance and to have remained static for hundreds of millions of years, forming Gondwana. But there is opposition to this view, some believing that a supercontinent prevailed from about 2.7 billion years ago until around 600 million years ago.

These seemingly wide and radically different interpretations of the geological and paleomagnetic data are entirely consistent with the precarious balance where slight modifications of analysis and interpretation can send alternative theories in opposing directions while each conforms to the known facts. Those scientists who believe that a single supercontinent existed for around 2.1 billion years point to recent interpretations of geology on other planets, particularly Mars and Venus where plates are known to have formed but where there is no movement or displacement. These 'lid' tectonics present a pancaked accumulation of lighter material forming on top of heavier and denser material with mobility. Plate tectonics, whether long or late in gestation, is unique to Earth.

Before passing from the Proterozoic eon to the Phanerozoic eon and events more recent than 541 million years ago, it is time to revisit the Wilson cycle and examine the five stages of continental development and evolution. The surface area of the continents has increased over time with parasitic accumulation of sediments and

micro-plates at the margins. It has taken several billions of years to build the continents to their present size, where 29% of Earth's surface is land mass and 71% water. Of the total continental mass, around 39% was formed in the Achean (2.5–4 billion years ago), 43% was formed during the Proterozoic (0.5–2.5 billion years ago) and only 18% in the Phanerozoic.

Wilson proposed, and it is now generally accepted, that convective cells from the asthenosphere ascend in what are referred to as diapirs, pushing up and through the lighter continental crust, creating faults, arches and ridges between which hot magma exudes and rapidly cools. This partial melting of the asthenosphere creates the alkaline-rich basalts mentioned earlier when discussing the Archean crustal formations. As the flow of magma separates either side of the ridge it thins out the continental crust through which it has appeared until it completely disappears. This continental material is replaced by ocean basalts that begin to form a new ocean, first by a small sea (such as the Red Sea) which is usually formed by water flowing in from an adjacent sea or ocean. Sometimes the flanks of the mid-ocean ridge grow to such proportions that they extend above the surface of the sea (Iceland, for instance) where they can be physically explored and examined.

The divergent motion of the oceanic crust, fed from the asthenosphere below, pushes the lighter continental plates in opposing, or rotating, directions, moving them around until, with increasing density, it begins to slip below the encountered continent through sea floor spreading. As its density exceeds that of the asthenosphere it sinks back down into the upper mantle at the subduction zone, dragging the rest of the ocean crust with it. In this process it can pull continents together, but instead of following it down into the mantle the approaching continental plate collides with the impacted land mass and buckles it, folds and causes uplift, building mountain chains and fractured rock seams causing earthquakes and settlement.

When the energy of impactor on the impacted continent is spent, erosion and weathering takes hold, reducing the mountain ranges and chamfering off the peaks and undulations. In this way, and by example, the great mountain chains

BELOW In snowball Earth theory, in which the whole Earth was covered in ice, the planet has passed through episodes where it has alternated between carbon-rich atmospheres of great pressure and heat to iceworld conditions. As the Sun's output has gradually increased those conditions have shifted to intermittent ice ages or partial glaciation. *(David Baker)*

that stretch from the Pyrenees across the Alps to the Caucasus, and the Himalaya itself, will eventually subside and become gently sloping hills and undulations on a continental scale. In this way the root and the antiroot described earlier come into isostatic equilibrium.

The Wilson cycle was a refined model for how Earth's engine works, but it is a machine slowly winding down, and just as the pace of production of continental materials peaked in the Proterozoic, so too will it eventually wither and cease to exist. But not for several billion years. Meanwhile, before leaving the Proterozoic eon, another major cycle of activity played a significant role in the way Earth developed. There is significant evidence to show that the Proterozoic was a period of active glaciation, resulting in what has been dubbed the 'snowball Earth' phase.

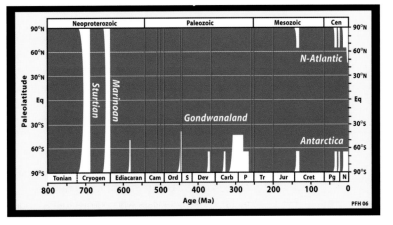

ABOVE Plotting glacial epochs against time and latitude from the equator to the poles, only before 630 million years ago were conditions appropriate for a global iceworld. The intensive glaciation of Gondwanaland around the Carboniferous and Permian boundary is because that supercontinent was at the South Pole during this period. (USGS)

Icehouse world

The onset of glaciation is, obviously, precipitated by a cooling phase, and as we have seen that can be caused by a variety of mechanisms, not least the reduction in the partial pressure of carbon dioxide which acts as an effective greenhouse gas maintaining temperatures higher than we have today, which is a period of relatively low CO_2. But the main regulator is methane, much more efficient than CO_2 at heating the atmosphere. Even the assembly of a supercontinent can cause such a cooling due to the weathering of silicates and precipitation trapping CO_2 in carbonates.

CO_2 and methane were essential components of the early Earth atmosphere to keep temperatures high enough to prevent water freezing and to establish and maintain an ecologically generous environment. Until quite recently in geologic time, it was only because of a greenhouse Earth that life was able to diversify, experiment and proliferate. But it also produced a trigger for the first major glaciation, itself precipitated by the onset of the Great Oxygenation Event (GOE). While atmospheric methane was significantly reduced and oxygen levels increased, many primitive life forms were made extinct and the Huronian glaciation arrived.

In several ways the geological and atmospheric effects of glaciation are linked,

but there are certain specific consequences that affect the Wilson cycle of plate tectonics in unexpected and dramatic ways. As Earth cooled rapidly around 2.4 billion years ago ice caps began to form at the north and south poles where there had been none. Continental landforms sitting close to the polar latitudes were temperate in comparison with their condition today – and had been for hundreds of millions of years. But as the transformation of Earth began from living organisms that changed the atmosphere, never again would the planet follow a barren path.

The end product of a glacial cycle, however long it lasts, is inevitable. Periods of intense

BELOW Stromatolites in the Australian National Park. A fossil form representing the growth of algal mats, they consist of concentric spherules or flat sheets of calcium carbonate and trapped silt and are encountered in limestone. (David Baker)

glaciation have inbuilt extinction factors in the way they trap carbon dioxide. With total glaciation solar energy is reflected back into space, adding to the cooling effect. But beneath the polar caps, growing ever larger and creeping down toward the equatorial belt, freezing everything out as it goes, the alteration of silicates cannot proceed and ceases, leaving the carbon dioxide exhaled by periodic volcanic activity to build up in the atmosphere once more. There are simply no remaining mechanisms to trap the greenhouse gas.

As the CO_2 builds, temperatures start to rise and the ice begins to melt. But as dry land appears once more the process of capturing CO_2 in carbonate rocks through extensive weathering, precipitation and water run-off starts anew. A sure indication of a recent glacial epoch is a major increase in carbon dioxide levels in carbonate formations. But there are changes to the nature of the subduction chemistry caused during a glaciation.

In warm times CO_2 is removed from the air by water in the form of carbonic acid (H_2CO_3),

weathering the rocks and carrying cations and bicarbonate ions (HCO_3^-) to the seas and oceans, where they get trapped in the carbonate rocks. The oceans normally process the calcium carbonate ($CaCO_3$) into sediments, and these get carried along the ocean floor into the subduction zone where they disappear into the mantle only to be regurgitated through volcanic events and mid-ocean ridges, where the whole process starts again.

The chemistry of the ocean floor changes during glaciation and this affects the way it reacts to naturally occurring chemical transformations. There is a very big difference between the action at the floor and at the subduction zone when the viscosity of the water changes in the solid state. Evidence of total ocean freeze-out is not difficult to find, and characteristic signatures of glacial movement on land can be seen on the sea floor where movement of the ocean crust moves against the ice (above) rather than the ice moving over the land (beneath).

The degree to which the glaciation ever became global in a literal sense is open to debate. Several sets of data indicate that there may have been an equatorial band around Earth that never completely froze over, leaving a slushy and listless section of water probably to within 10–20° of latitude, equating to a width of up to 1,300 miles (2,090km) between ice cliffs north and south, characterised by drifting ice floes interspersed with slush and icebergs. In this alien environment where temperatures in places could have dropped below -112°F (-80°C), it is difficult to imagine the conditions in a depleted atmosphere with low levels of carbon dioxide and methane but a relatively high proportion of oxygen.

Evidence from the late Archean rock formations in South Africa indicate there may have been a glaciation of considerably less magnitude 2.9 billion years ago, but with only trace indications as to extent. The Huronian glaciation which began 2.45 billion years ago was severe in the extreme, triggered by the Great Oxygenation Event. Evidence comes from what geologists call diamictites, sedimentary rocks comprising a wide mixture of fragments that are not formed in glaciers or ice conditions but which are modified by exposure to those conditions. When applied to true glacial environments they are referred to as tillite, and are found widely in North America and South Africa, the Huronian also associated with the major expansion of stromatolites – mats of microbacteria, including cyanobacteria (see Chapter 7).

Considerable debate surrounds the precise extent of the Huronian, and with evidence from more recent glaciations and a deeper understanding of the processes involved – not least from the more extensive evidential base for the Ice Age in which we exist today (see Chapter 6) – it appears that there were several such episodes between 2.4 billion and 2.1 billion years ago. This is a very broad span of time, and the early date of this event suggests there is still much to learn about how the icehouse world worked in the Proterozoic. But there is little doubt about a further spate of glaciation toward the end of this eon.

There is substantial evidence for a major glaciation between 900 million and 580 million years ago, sometimes known as the Cryogenian and agreed to as a formal period in 1990, now set by a fixed rock age of 720 million years. It appears that there were three periods of glaciation during this time, each lasting about 100 million years, centred on the Sturtian (715 million years), the Marinoan (635 million years), and the Varangian (580 million years). The extent of these glaciations is largely unknown but there are tantalising indications that the glacial front extended to quite low latitudes, certainly down as far as the Mediterranean or the North African coast on today's map. But the extent could have been very much greater and the great surprise with ice ages has been the general tendency to underestimate the extent of their advance as well as the totality of their effect. Again, some specialists question whether there was total global cover or only down as far as the latitude of Malaysia and Saudi Arabia today. Either way, it was a significant event.

Before we move to the Phanerozoic it is worth pondering the almost constant infall of debris, including asteroids and meteorites left over from the accretion process, the readjustment of the orbits of the outer giants which brought the Late Heavy Bombardment

and the general chaos caused by what scientists today call Near-Earth Objects (NEOs) impacting Earth – occasionally with dramatic consequences and occasionally with catastrophic results.

Only 179 major meteorite impacts have been identified to date, compared with several tens of thousands on the Moon. Plate tectonics, weathering and the general transformation of the surface through mountain building and domical rounding of those peaks caused by erosion have all but obliterated evidence of their former existence. Because the oldest ocean crust is aged at no more than 180 million years, a lot of impact events could have occurred only for the evidence to have been sent down into the mantle at the subduction zones. None of the known impact events date back to the Archean, the majority being in the Proterozoic from 2.5 billion years to a mere 541 million years.

The two which stand out most are the Vredefort crater in South Africa with an age of 2.023 billion years and a diameter of 186 miles (300km) in its original context. The impact was so severe it substantially modified the Witwatersrand basin and the Ventersdorp lava fields, which had been laid down between 2.95 billion and 2.70 billion years ago. Estimates of the energy released to create this giant crater indicate that the meteorite must have been 3.1–6.2 miles (5–10km) in size. There is one older meteorite impact, the Suavjärvui crater in Karelia, Russia, with a diameter of 10 miles

ABOVE The world famous Meteor Crater in Arizona, spectacular evidence of impacts that have periodically caused major changes to the Earth and its atmosphere. With a diameter of 0.74 miles (1.19km), it is believed to have been caused by a meteor 160ft (50m) across that hit the Earth about 50,000 years ago.
(David Baker)

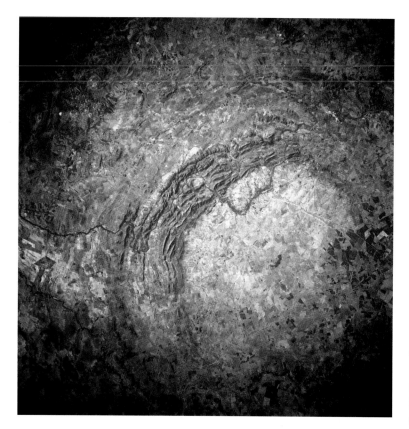

which is not found on Earth, iridium being one and a sure indication of a meteorite impact.

More recent impacts from the Phanerozoic display droplets of molten rock ejected during the impact, rising high into the atmosphere and solidifying before they fall back to the ground far away from their point of origin. Sometimes these droplets, while mainly spherical in shape, form aerodynamic shapes as they enter the denser layers of the atmosphere. They may have been out into space before falling back, or they may achieve escape velocity from the force of the impact, reaching a speed of more than 17,500mph (10,875kph) to enter an elliptical orbit about Earth. Some will reach escape velocity of 25,000mph (15,535kph) to leave Earth for ever and wander around in the solar system until drawn by gravitational attraction to another planet.

This is not an irregular occurrence, at least on the span of geologic time, because ejected material from other planets experiencing similar impact events has landed on Earth, uniquely identifiable with the isotopic signatures that declares their origin, material that undoubtedly came from the Moon and Mars. Like Earth, both bodies continue to encounter random impacts, and cometary debris is responsible for some of that. There are even commercial companies which sell pendants allegedly containing powdered Moon rock from ejecta collected on Earth. Buyer beware!

Earth today

This is the age of the Phanerozoic, the last of the four eons dividing Earth into separate phases of its evolution, largely dictated by the presence, and proliferation, of life. As we shall see in Chapter 7, life has been present on the planet almost from the beginning, but this last eon was set down as a marker for life as we know it today and the onset of abundant flora and fauna, the first fossilised examples of that life and the development of intelligent animals, allegedly including humans.

The Phanerozoic is the story of the last 541 million years. Everything that came before is the Precambrian. It is divided into three eras: the Paleozoic; the Mesozoic; and the Cenozoic. Those eras are further subdivided into 12 periods: Cambrian, Ordovocian, Silurian,

ABOVE Of major significance, the Vredefort basin in South Africa is typical of the giant impact events that occur throughout the history of the Earth. Seen from a Shuttle Orbiter in its entirety, the basin is 186 miles (300km) across and was created around 2 billion years ago in an impact that remodelled the Witwatersrand basin and associated lava fields. *(NASA)*

(16km) and judged to be 2.4 billion years old, but very little survives today, only surface shock features and some breccias lying around.

The second interesting meteorite impact is in the Sudbury basin, Ontario, itself 39 miles (62km) wide and 19 miles (31km) long with a depth of 9.3 miles (15km). The impacting meteorite must have been 6.2–9.3 miles (10–15km) across because the energy formed a crater 160 miles (250km) in diameter that threw its ejecta blanket a distance of more than 500 miles (800km), sending ejecta rays right around Earth. Rock fragments from the impact have shown up in geological surveys in Minnesota, 600 miles (967km) away.

Evidence of meteorite impacts comes from a variety of sources and range of rock and mineral types. As we have seen in Chapter 2, the early Earth differed in several respects from the material that made up the asteroids and meteorites. They have distinct geochemical signatures that tell a story with conviction because they are unique to certain types of object in the solar system. Geologists find nickel-rich spinels, shocked minerals and impact spherules. There may, occasionally, be the telltale signature of a chemical or an element

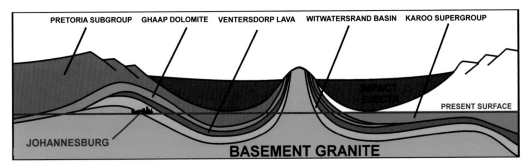

PRETORIA SUBGROUP   GHAAP DOLOMITE   VENTERSDORP LAVA   WITWATERSRAND BASIN   KAROO SUPERGROUP

IMPACT EJECTA

PRESENT SURFACE

JOHANNESBURG

BASEMENT GRANITE

LEFT A cross section of the Vredefort basin reveals the upthrust which has been eroded in the eons of time since impact, with Johannesburg shown for location and scale. *(David Baker)*

Devonian, Carboniferous, Permian, Triassic, Jurassic, Cretaceous, Paleogene, Neogene and Quarternary. Each characterises a specific set of boundaries involving the geology of Earth and of its living systems. There are further subdivisions into ages but those are highly specific and can muddy the generalised picture that we are viewing here.

The Paleozoic runs for 289 million years, from 541 million to 252 million years ago. It is the longest span of Phaneorozoic time and is characterised by the laying down of many of the materials that power human occupation, communities and cities, including coal and fossil fuels. This era also saw a major mass extinction of life across all biota, flora and fauna.

It is succeeded by the Mesozoic, which lasted for 186 million years, stretching from 252 million to 66 million years ago. This era saw the rise of one of the more successful forms of land animal – the dinosaur and the branch that took to the air with feathers and became birds, survivors to the present. But it also ends with another great extinction, that of the non-avian dinosaurs being the most well known.

Finally, the Cenozoic covers the last 66 million years in which the world we know today grew and took form, opening niches for the rise of the mammals and the primates and, in the very last sliver of time, the emergence of humans. It is a period dominated by the emergence of animals and plants after the cataclysm of the great extinction at the boundary of the Cretaceous and the Paleogene.

To put this in scale, if the entire history of Earth to the present time were represented by a calendar year of 365 days, the 541 million years of the Phanerozoic would occupy the last six weeks. The dinosaurs would evolve at the end of the first week in December and die out in a catastrophic event on 26 December. Primates would arrive just after midnight at the beginning of

the last day and early hominids would walk Earth from 4:00pm as evening was approaching.

Modern humans like you and I would have evolved a mere 20 minutes before midnight, sharing the planet with Neanderthals and probably the last surviving members of other hominid strains. Only in the last 60 seconds of the year would writing, agriculture and the first cities appear, and only 20 seconds before the end of the year would Stonehenge get built, monotheistic religions appearing within the last 10 seconds. While seemingly a fleeting whisker of time, the Phanerozoic being only the last six weeks of a year on the scale of Earth's total story, much happened to transform the planet into a platform for burgeoning life, that story being taken up in Chapter 7.

The geological story of Earth in this eon incorporates the last known assembly of continental plates into the supercontinent from which our present continents have emerged and from which others are still forming. Geological events of the past 541 million years focus on this single event as probably the most important for the way it interacts with the evolution of living organisms, while

BELOW An evolution of the Earth's radiogenic energy output from the different decay rates of uranium, potassium and thorium showing how the total thermal input to the mantle is reducing with time. *(David Baker)*

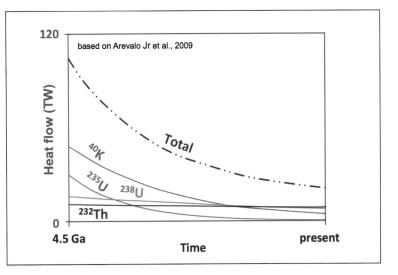

interaction between the physical action of volcanic eruptions and the atmosphere is represented by the traps (thick piles of volcanic rock) which contain the record of vast swathes of land covered with basaltic lavas in episodes stretching over tens of thousands of years on various continents across the globe.

At the boundary of the late Paleozoic and the early Mesozoic, around 250 million years ago, the great supercontinent of Pangea was formed from two clusters of continental masses known as Gondwana and Laurasia. It had its origins 500 million years earlier in the Phanerozoic when Rodinia broke up and separated into three pieces – a cluster known as Proto-Gondwana, another cluster known as Proto-Laurasia and the Congolese craton. They were separated by a tract of water known as the Proto-Tethys Ocean. Shortly after this, Proto-Laurasia itself split into three separate plates to become the continents of Laurentia, Siberia and Baltica. Baltica moved east of Laurentia and Siberia moved to the north of Laurentia, creating two new oceans, the Iapatus and the Paloasian.

As a result of the separation and drifting the continents converged again to form the supercontinent of Pannotia, relatively short-lived before separating again around

540 million years ago, right at the beginning of the Paleozoic. But as the Cambrian period (542 million to 485 million years) began the brief encounter split, and from this renewed drifting arose Laurentia, Baltica and Gondwana. Laurentia would become North America and was located on the equator with the Panthalassic Ocean to the north and the west, the Iapatus to the south and the Khanty Ocean to the east.

About 480 million years ago, at the start of the Ordovician (485 million to 440 million years) a tiny continent, Avalonia, formed which comprised an expanse of land eventually to become eastern Newfoundland, the southern British Isles, parts of Belgium and northern France, Nova Scotia, New England and north-west Africa. Breaking free from Gondwana it drifted across to Laurentia, which, together with Baltica, came together at the end of the Ordovician to become the supercontinent Euramerica, bordered by the Iapatus Ocean. The impact of this collision formed the Appalachian Mountains. Siberia was situated alongside Euramerica, separated from Gondwana by the Khanty Ocean.

Meanwhile, with Gondwana drifting toward the South Pole, Euramerica basked at the equator, the expansive rain forests of the

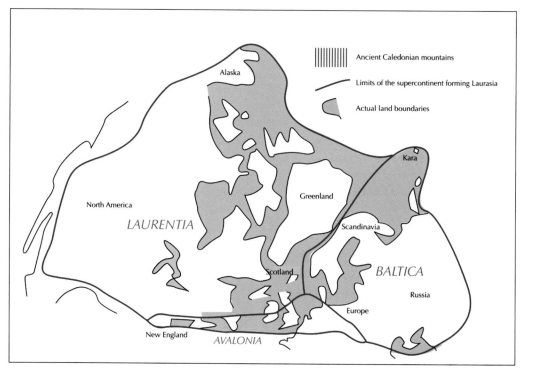

RIGHT Destined to join up together as Gondwana, the land masses of Laurentia, Baltica and Avalonia converged and joined up about 400 million years ago, immediately prior to which Euramerica, comprising Laurentia and Baltica, had formed. Avalonia would bequeath part of its land mass to what would become southern England, northern England and Scotland being part of Laurentia.
(David Baker)

Ancient Caledonian mountains

Limits of the supercontinent forming Laurasia

Actual land boundaries

Alaska

Kara

Greenland

North America

LAURENTIA

Scandinavia

Scotland

BALTICA

Russia

Europe

New England

AVALONIA

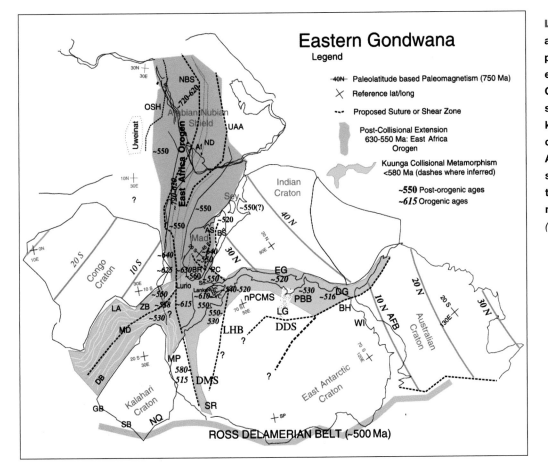

Eastern Gondwana

Legend

-40N- Paleolatitude based Paleomagnetism (750 Ma)

✕ Reference lat/long

--- Proposed Suture or Shear Zone

Post-Collisional Extension
630-550 Ma: East Africa
Orogen

Kuunga Collisional Metamorphism
<580 Ma (dashes where inferred)

~550 Post-orogenic ages
~615 Orogenic ages

ROSS DELAMERIAN BELT (~500 Ma)

Carboniferous (359 million to 299 million years) laying down vast resources for humans to scavenge 320 million years later. However, the collapse of the rain forest around 305 million years ago was probably precipitated by the separation of large land areas into small islands on the continental plate, which changed the coal-bearing seams. While rain forests would continue to feature in tropical zones, their location would change and never return to the vast swathes seen in the Carboniferous.

This the first phase in the migration of continents toward Pangea, another being the inevitable convergence of Gondwana and Euramerica. Long before that, at the start of the Silurian 440 million years ago, Baltica collided with Laurentia to form Euramerica but Avalonia had yet to join on. Although a small plate totally covered with water, Southern Europe separated from Gondwana and moved toward Euramerica across a newly formed seaway named the Rheic Ocean, colliding with the southern flank of Baltica in the Devonian (419 million to 359 million years). The Khanty Ocean then began to shrink when

an island arc from Siberia struck eastern Baltica, which had now become part of Euramerica.

By the late Silurian, around 415 million years ago, areas today known as north and south China separated from Gondwana and began moving northward, reducing the width of the Proto-Tethys Ocean, opening what became the Paleo-Tethys in the south. During the Devonian period the land mass of Gondwana began to move toward Euramerica, shrinking the Rheic

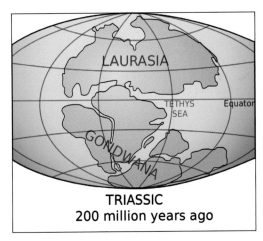

TRIASSIC
200 million years ago

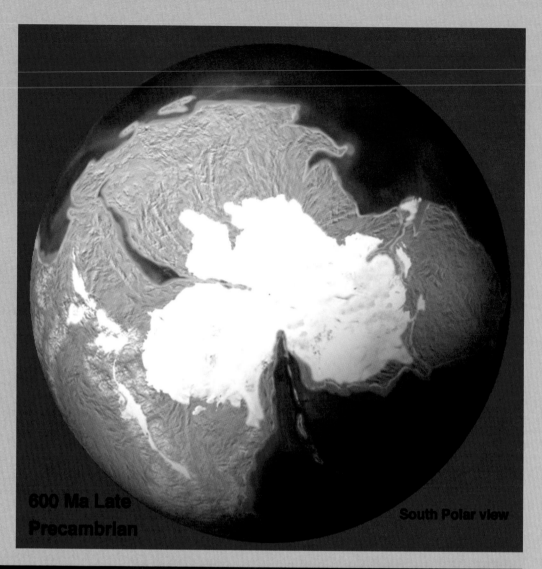

RIGHT Late
Precambrian
600 million years ago.
(Greatarsson)

BELOW Cambrian
500 million years ago.
(Greatarsson)

600 Ma Late
Precambrian

South Polar view

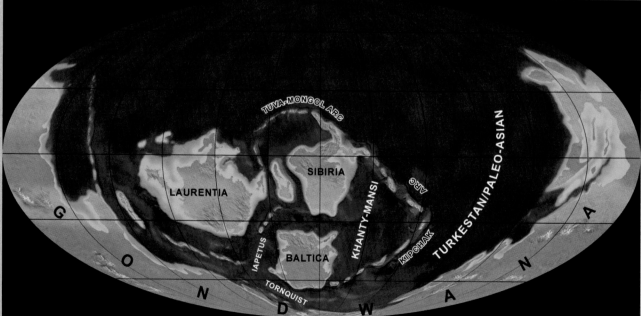

TUVA-MONGOL ARC

SIBIRIA

LAURENTIA

KHANTY-MANSI

ARC

KIPCHAK

TURKESTAN/PALEO-ASIAN

G

O

N

D

W

A

N

A

IAPETUS

BALTICA

TORNQUIST

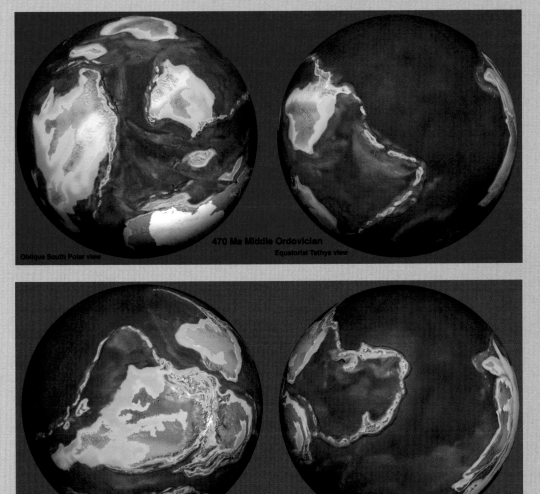

470 Ma Middle Ordovician
Oblique South Polar view   Equatorial Tethys view

LEFT Middle Ordovician 470 million years ago. *(Greatarsson)*

430 Ma Early Silurian
Equatorial West view   Equatorial East view

LEFT Early Silurian 430 million years ago. *(Greatarsson)*

BELOW Early Devonian 400 million years ago. *(Greatarsson)*

TUVA-MONGOL ARC
SIBIRIA
KHANTY-MANSI
KASAKHSTAN OROGCLINE
TURKESTAN/PALEO-ASIAN
LAURUSSIA
PALEO-TETHYS
G O N D W A N A

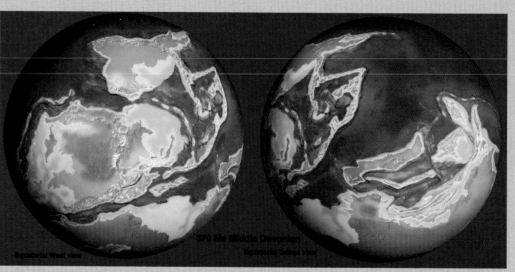

RIGHT Middle Devonian 370 million years ago. *(Greatarsson)*

Equatorial West view

370 Ma Middle Devonian

Equatorial Tethys view

BELOW Middle Devonian 380 million years ago. *(Greatarsson)*

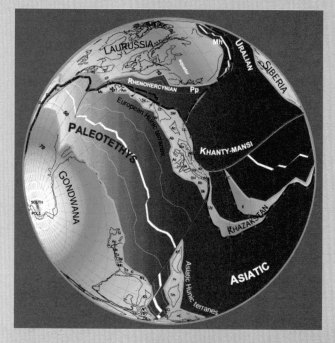

LAURUSSIA

Mh URALIAN

SIBERIA

Rhenohercynian

Pp

European Hunic terranes

PALEOTETHYS

KHANTY-MANSI

GONDWANA

SOUTH POLE

KHAZAKSTAN

Asiatic Hunic terranes

ASIATIC

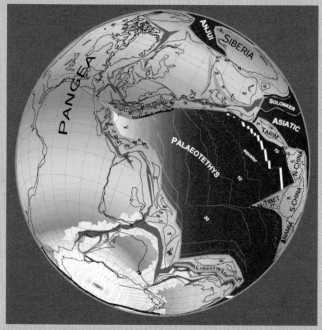

ANJUI SIBERIA

PANGEA

SOLONKER

ASIATIC

TARIM

PALAEOTETHYS

N-CHINA

N-TIBET

S-CHINA

ANAMIA

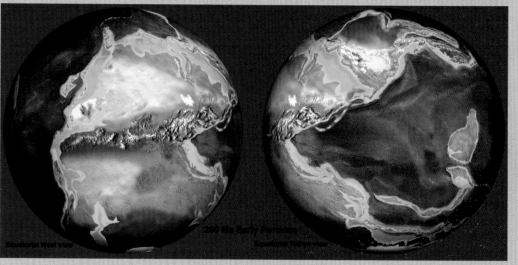

RIGHT Early Permian 280 million years ago. *(Greatarsson)*

ABOVE RIGHT Early Permian showing Paleotethys. *(Greatarsson)*

Equatorial West view

280 Ma Early Permian

Equatorial Tethys view

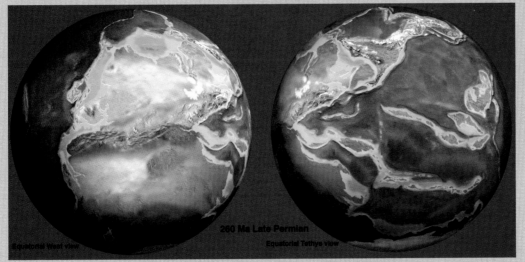

260 Ma Late Permian

Equatorial West view   Equatorial Tethys view

LEFT Late Permian
260 million years ago.
(Greatarsson)

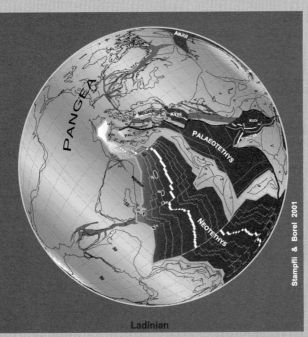

Ladinian

LEFT Upper Triassic
230 million years ago.
(Greatarsson)

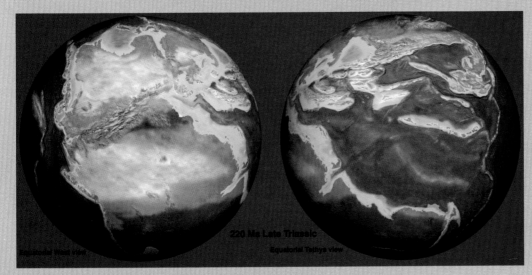

220 Ma Late Triassic

Equatorial West view   Equatorial Tethys view

LEFT Late Triassic
220 million years ago.
(Greatarsson)

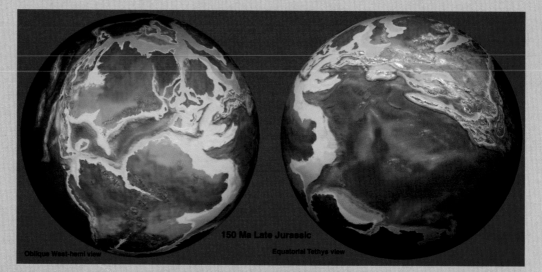

150 Ma Late Jurassic

Oblique West-hemi view

Equatorial Tethys view

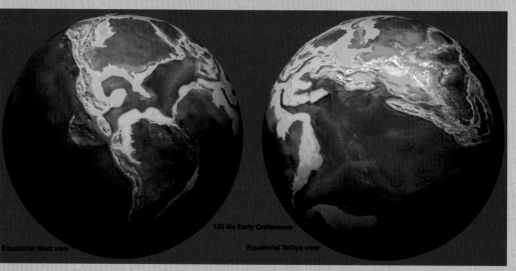

120 Ma Early Cretaceous

Equatorial West view

Equatorial Tethys view

105 Ma "Middle" Cretaceous

Oblique West-hemi view

Equatorial Tethys view

Ocean in the process, and by the time of the early Carboniferous, around 350 million years ago, north-west Africa was in juxtaposition with the south-east coast of Euramerica. This created the southern portion of the Appalachian Mountains while South America moved north to arrive at southern Euramerica.

By this time the eastern section of Gondwana, comprising what are today India, Antarctica and Australia, was still heading toward the South Pole on its way from the equator. Although it had been on its own for several million years, Siberia was joined by the Kazakhstania microcontinent, the western side colliding with Baltica about 320 million years ago, closing the western Proto-Tethys and forming the Ural Mountains. This movement eventually brought about the creation of the Laurasia supercontinent, but in the meantime South America struck southern Laurentia, closing the Rheic Ocean completely, setting up the southern part of the Appalachians.

Gondwana was now at the South Pole and glaciers and ice sheets were forming across Antarctica, southern Africa, South America, Australia and India. As the north China slab struck Siberia 310 million years ago, it shut off the Proto-Tethys Ocean. Early in the Permian period (299 million to 252 million) a sub-plate of the Gondwana land mass split away and completely closed the Paleo-Tethys Ocean but in the process formed a new Tethys Ocean at its southern end. The land masses were now united as one supercontinent locked together in the southern hemisphere and wrapped around the equatorial zone. The general shape of Pangea was in the form of a crescent with the Tethys Ocean in the bowl of the curve.

The evidence for the existence of Pangea and for the last great supercontinent, until the next is formed several tens of millions of years from now, comes not only from the geologic record but from the prolific array of fossils and both flora and fauna which substantiate the record from the rocks. Age-dating is a key to putting together the story of this mass migration of land masses into one agglomeration, which lasted for around 25 million years until a sequence of events began to disturb the configuration and start the process of separation. Mountain chains bear the evidence of juxtaposition, the alignment of the Appalachian and Caledonian mountains for example which were once a continuous chain but are now on opposite sides of the Atlantic Ocean. The arrangement and wander paths for the various components of Pangea, as outlined above, can be mapped not only through evidence from age-dating and from the folds and faults that accompanied these continental movements: it is also found in the record of paleo-magnetism.

The way the poles have appeared to move all around the surface of Earth, oblivious to the rotation of the planet, testifies not to the actual shifting of the magnetic axis – although that does change a little over time, as we have seen – but rather to the movement of the continental slabs that record the magnetic polarity at different times. Scientists can measure quite precisely the orientation of magnetic minerals when the rocks were formed in their present state, either by igneous or metamorphic processes, and subtract the natural wandering of the magnetic pole. The misnamed 'polar wandering' is in fact the wandering movement of the continental plates, which show where they were at different stages in their migration and not the large-scale movement of the magnetic pole itself.

Beginning in the first half of the Jurassic period (which extended from 201–145 million years), about 180 million years ago the Tethys Ocean in the east began to rift from the Pacific region in the west, creating Laurasia and Gondwana in a lengthy and protracted sequence in which the old Gondwana – comprising what is now South America, Africa, Antarctica, India and Australasia – separated from Laurasia in the northern half of the Pangea supercontinent. Massive basaltic eruptions preceded the migration of this land mass. Rotating away and severing from Laurasia, the collective mass of Antarctica, Madagascar, India and Australia broke off from Africa. South America moved westward from Africa and a new ocean, the South Atlantic, began to open up around 130 million years ago. Within 20 million years there was open water between the two, by which time India had begun its journey northward.

Meanwhile, farther south, Antarctica lost a piece of its continent 85–100 million years ago when what is now New Zealand separated. Australia also started a long and protracted period of separation 80 million years ago, and to move east toward the equator. As it moved at a fast pace north, a slab broke away from India to become Madagascar and a small sliver, which became the Seychelles, translated north too. This event occurred right at the time of the great mass extinction at the end of the Mesozoic 66 million years ago, apparently finishing off the dinosaurs and many other species. This was also when the Deccan Traps were laid down, one of the most violent events in the Phanerozoic.

Located in west-central India, the Deccan traps represent one of the most fascinating geological features on Earth for geologists studying the expansive piles of basaltic lava on continental crust. Traps are probably connected directly to plumes from the outer mantle that penetrate continental crust and overlay the lighter silicaceous material with vast pancakes of lava, creating large areas with climatological influence far beyond their size relative to the planet. On a continental scale they are, however, very large, and the Deccan traps is one of the largest of several such features on Earth. Today, it covers an area of 193,000 square miles (499,900km²) with a total volume of 123,000 cubic miles (512,000km³). At its formation it would have been much larger, releasing a calculated volume in excess of 264,000 million cubic miles (1.1 million km³) but erosion has reduced its visible area.

The traps were influential in the great extinction that occurred at the end of the Cretaceous 66 million years ago, as they had been in previous extinction events. Most notable in the preceding mass extinction was the Great Siberian Trap at the boundary of the Permian and Triassic, around 250 million years ago, which laid down an expanse of basaltic lava up to 3.5km³ million in volume. It could have been either the direct cause of the Permian extinction or itself a triggered event from a giant meteorite impact which compounded the destruction; either way traps are notorious for being around when great convulsions in the balance of climate and life take hold of Earth.

The Deccan traps were at their worst in the region of the Western Ghats close to the modern city of Mumbai, overall a series of eruptions lasting probably 750,000 years. During and immediately after these extrusions the temperature of Earth fell as the atmospheric changes obscured sunlight. Very precise dating of the ten primary formations that make up the Deccan traps show that it was at its peak a mere 25,000 years before the great Cretaceous-Paleogene extinction and that it lasted for around 753,000 years. Whatever the cause, this and similar traps which are consistently found to have occurred at irregular intervals may have the same physical trigger as that which determined the extruded lava beds characterised by long lobate flow ridges on the near side of the Moon – mantle plumes.

While basalts extrude through mid-ocean ridges and lay down ocean crust, material from mantle plumes that form traps on the surface have subtle differences in their chemical composition, containing slabs of splintered continental sediment and ocean crust subducted into the upper mantle. The very subduction process separates the water-soluble trace elements such as potassium, rubidium and thorium from the immobile trace elements titanium, niobium and tantalum, concentrating them in the oceanic slabs while the water-soluble elements get attached to the crust in island arcs. These oceanic slabs descend to the boundary of the mantle transition zone at a depth of about 400 miles (650km), some perhaps sinking down as low as 930 miles (1,500km), and this prepares the base of material that the plumes transport to the surface.

BELOW The Deccan Traps, notorious for having dramatically changed the very nature of the atmosphere, stark reminder that a single, comparatively moderate event in Earth's history can, and will continue to directly affect climate, life and the future evolution of the atmosphere and oceans. The expansive volcanic lava extruded at the Deccan Traps was probably more responsible for the mass extinction at the end of the Cretaceous than the giant impact event at Chicxulub off the Yucatan Penninsular, Mexico. *(David Baker)*

The temperatures at Earth's core are 1,832°F (1,000°C) higher than those of the adjacent mantle, 1,865 miles (3,000km) below the surface, and there is very little transport of material between the two, which is why thermal transfer takes place through conduction. It is believed that as the base of the mantle becomes hotter it gives rise to detached plumes that, on rising through the mantle, become hotter still and more buoyant. As it ascends even further toward the surface it enters a phase known as decompression melting caused by a reduction in pressure, and this allows giant plumes of lava to form which collect the chemical discontinuities mentioned earlier and project them through the continental crust as traps.

The physical principles behind the thermal plumes, or 'hot spots' in the mantle, explains the way island chains of volcanic material form, a vertical plume from the lower mantle travelling up through a moving plate at periodic intervals, penetrating the surface to create a string of volcanic domes, the Hawaiian islands being a classic example. In seismological studies of Earth the core-mantle boundary is known as the D" layer, and it is here that the evidence for plume origin has been explored using seismological tools. It appears that there may be two enormous super-plumes permanently attached to the D" layer from which separated plumes depart and eventually reach the surface.

ABOVE The Siberian Traps released vast quantities of carbon dioxide into the atmosphere. Lava was laid down across vast areas, including here at Puorana. *(David Baker)*

The Hawaiian-Emperor seamount chain, as it is known, consists of 129 volcanoes across a 3,600-mile (5,800km) range rising 15,000ft (4,572m) above the sea floor. Accumulated lava flows grow at a slow rate, creating shallow-angle slopes growing from east to west, a sequence of events that began around 81 million years ago but which underwent a rifting 43 million years ago in a process which continues today, the youngest volcano being Kaua'i, which is a mere 5 million years old. Some of the islands have gone through the coral-island forming stage where the volcano becomes dormant and the centre collapses,

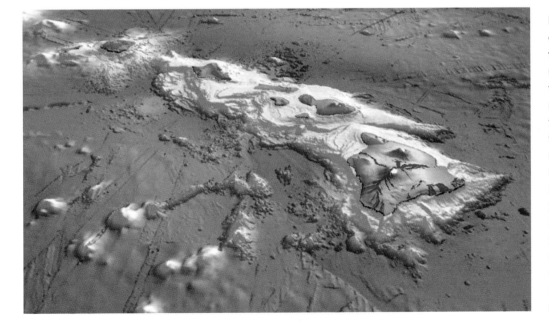

LEFT As explored earlier in this book, the movement of the ocean crust across the asthenosphere allows venting from thermal mantle plumes to puncture the surface at a serious of intervals, never building up a single massive volcano, but liberating the energy through a series of smaller island volcanoes, as seen here with the Hawaiian seamount. *(NOAA)*

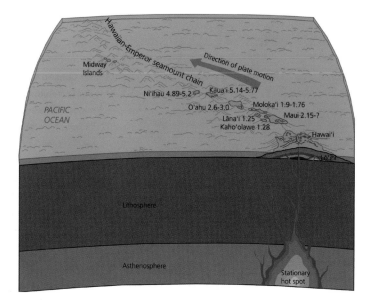

ABOVE The direction of plate motion can be charted using the island chains as punctuation marks in time, the drift rate being shown here in cm/yr of motion. *(USGS)*

BELOW Lava sheets pouring out to the ocean at the Pahoehoe site in Hawaii. *(USGS)*

its content subsiding to leave a ring, or atoll, and a coral reef where once a mighty volcano stood above the sea. Further subsidence or total collapse of the walls can produce a seamount, where all that remains are sloping undulations only vaguely reminiscent of its origin, submerged and only seen by divers.

The Deccan traps of 66 million years ago and the Siberian shield of 250 million years ago are but two examples of the impact ascending thermal plumes have had on the evolution of Earth's surface. The same processes that caused the conveyor-belts of plate tectonics

drive the surficial changes that have moulded Earth for billions of years. The same physical principles link thermal upwelling at mid-ocean ridges to ocean volcanoes and island chains to the formation of vast expanses of basaltic lava flows that have changed considerably the course and evolution of both the climate and living organisms.

During the Mesozoic, the era bridging these two cataclysmic events and in some way defined by them, there were many geological incidents that show strong indications of connection between infalling objects and induced geologic activity. The apparently random giant meteorite impacts appear strongly linked to these episodes of traps and exuded plumes, but possibly the most well-known is the Chicxulub event at the end of the Cretaceous 66 million years ago, contemporaneous with the major extrusions at the Deccan traps. The event has achieved global fame in popular literature as the hypothesised cause of the non-avian dinosaur extinction for which a cause had been sought for generations.

The fossil record shows that around 66 million years ago a broad swathe of land and marine life became extinct, as it had at the Permian extinction 184 million years earlier. Environmental change was brought about on a global scale. The associated impact at Chicxulub formed a crater 6.2 miles (10km) in diameter, releasing 10 billion times the energy released by the atomic bomb dropped on Hiroshima. This single impact at the end of the Cretaceous was 400 times the eruptive yield of the largest volcano known, the La Garita event in Colorado when a series of massive explosions 26–28 million years ago, liberating 1,200 cubic miles (5,000km³) of pyroclastic ignimbrite, spewed out so much material that it would have been sufficient to fill Lake Michigan.

The Chicxulub impact zone is an area off the coast of the Yucatan peninsula in Mexico, and the impactor penetrated marl and limestone that covered the area to a depth of 3,300ft (1,000m), below which were a layer of breccias and andesite. The glass andesitic rocks, formed under shock along with quartz, are found within the crater side, part of which is offshore, while the crater itself penetrated down 3,600ft (1,100m), its gently sloping walls

reaching a depth of 1,600ft (500m) about 3 miles (5km) from the centre. Underground there are numerous caverns and caves formed from sinkholes immediately after the impact, the effects of which were felt around the world.

It would have taken less than a second for the incoming meteorite to have penetrated to maximum depth, the impactor heated to incandescence, glowing white hot and with a plasma shock wave trailing behind, heating the atmosphere as it went. Material thrown out and up from the impact would have been hurled far out into space, gradually returning over several hours bringing a rain of fiery death, lasting from several hours to possibly several days after the event reminiscent of the Late Heavy Bombardment. Fires and scorched earth would have been ignited across the globe by dust and ash from the impact falling back to the ground. Colossal shock waves raced around the planet, colliding on the opposite side of the globe and causing folds and chaotic ruptures where they met.

Dust and fine particles would have covered much of the planet for several decades and the Sun would have been shut out for months. Stress on life would have been extreme, with massive release of carbon dioxide from shocked and shattered carbonate rocks raising the temperature of Earth through an induced greenhouse effect which would see an

Trough

Cenotes
(sinkholes)

alarmingly steep rise in CO_2 levels on a scale impossible to create artificially. While the long-term effects were self-restoring, short-term the reduction in photosynthesis would have severely restricted plant life, and on an even shorter term the tsunami caused by the impact, largely in water, would have devastated vast areas of land inundated with tidal waves up to 1,000ft (300m) in height, intruding upon vast areas of continental land.

We will revisit the climatological effects of these, and other giant impact events, in Chapter 6, but the progression of continental

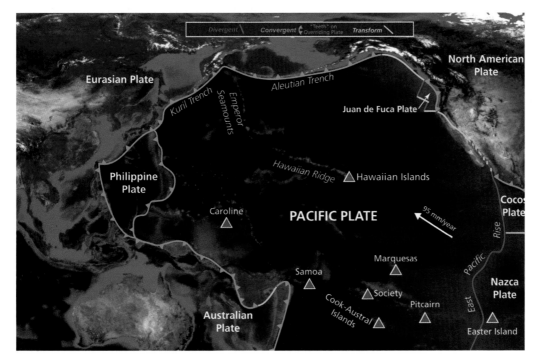

ABOVE The Santa Ana volcano in El Salvador is a cinder cone volcano at an elevation of 7,800ft (2,380m). *(USAF)*

dispersion punctured by the Chicxulub impact was only a temporary blip. Some 10 million years after the great extinction, about 55 million years ago, Australia and New Guinea started to separate and rotate northward, and within 10 million years the southern tip of Australia began to separate from Antarctica and form Tasmania with ocean flowing between the two continents for the first time.

It was at this time, 45 million years ago, that India began burying itself into Asia, buckling and folding the crust to begin forming the Himalaya and to create folds far to the north in areas that are now Tibet and Nepal. The lengthy attachment of Australia to Antarctica had kept the latter relatively warm. Ocean currents flowing up into the subtropics and around the

gradually rotating Australia, still attached, had kept Antarctica basking in a relatively balmy environment. When Australia finally separated and Antarctica was encircled by ocean its connection with warmer climes was broken.

South America separated from west Antarctica about 30 million years ago and this too caused great climatic changes, which had a ripple effect across the whole planet as major alterations to atmospheric and oceanic circulatory forces had their impact on the weather systems throughout the southern hemisphere. Although in theory South America had been attached to Antarctica until this time, it was in fact separated but linked by several micro-plates that allowed transfer of biological matter between the two. As ice began to build up on Antarctica the continent cooled and the temperature of the sea fell by about 18°F (10°C).

To the north-east, that edge of the Australasian plate collided with the south-west section of the Pacific plate buckling the land form of New Guinea, the extended arm of the Australian continent. This caused the New Guinea Highlands to form, eventually lifting the highest mountain in the central cordillera to 13,248ft (4,038m). While creating a fertile country it served as a highly effective rain shield and thus began the drying out of Australia, causing the expanses of desert and arid terrain with which Australia is blighted today in all but its coastal regions. At present both Australia and India are migrating north at a rate of 2–3in per year (5–6cm per year), and while India is burying itself into Asia, Australia will eventually collide with eastern Asia.

Today, Earth is subject to exactly these same forces, but the clarity of our perception is blinkered by the extraordinarily brief period in which we, as humans, have been around to absorb the fact that these realities are around us, operating on an extremely slow (geologic) time scale. Right now, Earth is covered with volcanoes, earthquake zones and faulting which are far beyond our control and which we are oblivious to on the day-to-day scale of ordinary life – until a major catastrophe caused by an eruption, a quake or a tsunami jolt us from our blinkered complacency.

Volcanoes come in different forms, types and sizes and the majority of volcanic activity is happening on a continuous basis, spreading

RIGHT Stratovolcanoes are arguably the most dramatic volcanoes, ejecting lava and other material to create a series of layers, or strata. Typical are Mounts Vesuvius and Stromboli in Italy, Mt Fuji in Japan and Mayon in the Philippines. Key: 1, large magma chamber; 2, bedrock; 3, conduit; 4, base; 5, sill; 6, dyke; 7, ash layers; 8, flank; 9, layers of emitted lava; 10, throat; 11, parasitic cone; 12, lava flow; 13, vent; 14, crater; 15, ash cloud *(Messer Woland)*

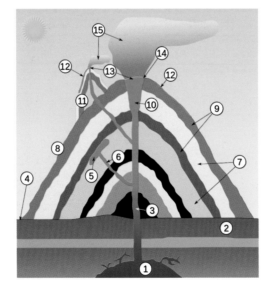

RIGHT One of the world's more famous volcanoes, Mount Vesuvius as seen from Naples seems passive and beautiful in the clear sky. As a stratovolcano it is anything but calm when it erupts, forcing vast quantities of lava through an explosive vent tube. *(David Baker)*

the sea floor, moving the continents around and laying down conveyor belts of cooling magma that flows far beneath the oceans we sail across. Divergent plate boundaries are fed by volcanism, black smokers revealing deep ocean vents where the surface of Earth is produced. Occasionally we get to walk around on some spectacular examples of divergent boundaries where the ridges are so tall that they protrude above the sea. Such a place is Iceland.

As we have seen, thermal plumes also bring volcanic products on to the surface of Earth, and convergent plate boundaries produce subduction zones that induce stress, heat and thermal loads, all of which result in both volcanoes and earthquakes. One kind of relatively benign eruption is through fissure vents, or lava tubes, where magma flows without explosive discharge, and these are found in many places, such as Iceland where the American and Eurasian plates are diverging. These can sometimes be accompanied with explosive eruptions at fissure intersections, and historic fissure can also be found on some Atlantic islands such as Lanzarote, where the last flow occurred in 1824.

Other major archetypical volcanoes include shield, stratovolcanoes, lava domes and cinder cones. The shield volcanoes differ distinctively in that they have highly mobile magmatic flow producing broad sloping flanks in a series of almost continuous eruptions. The largest on Earth today is the Hawaiian Mauna Loa, which rises 13,678ft (4,169m) above sea level and is more than 60 miles (97km) wide at its base and is believed to contain about 19,000 cubic miles (80,000km³) of basalt. The total height from its base on the ocean crust is 56,000ft (17,170m), which is about 27,000ft (8,230m) higher than Mount Everest is above sea level. Obviously these are anomalous comparisons; the real measure lies in its comparison with the largest known volcano in the solar system.

On Mars, where plate tectonics has produced separate static slabs of crust, there is no lateral movement to prevent mantle plumes from building up to truly gigantic volcanoes. The same physical principles that govern geology on Earth are to be found on other worlds throughout the solar system, and comparative planetology, as it is known, adds measurably to an understanding of not only our Earth but of other planets as well. Unlike the Hawaiian island chain, moving across the surface of the mantle, on Mars the largest volcano stands across an area almost equivalent to the land area of France. Named Olympus Mons (Mount Olympus), it stands 69,650ft (21,230m) above the local plateaus but 85,000ft (26km) above

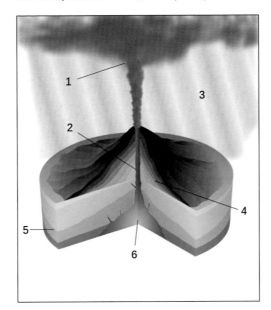

LEFT Volcanoes come in a wide variety of types and structural arrangements. So-called Plinian eruptions are spectacular and are usually caused by pressure at subduction zones where volcanic arcs are common. Key: 1, ash plume; 2, magma conduit; 3, volcanic ash rain; 4, separate layers of lava and ash, 5: stratum, 6: magma chamber. *(David Baker)*

RIGHT A typical Plinian eruption as a volcano on the Kenai penninsula lights up. *(David Baker)*

the plains that surround it. It has a base diameter of 370 miles (600km).

Olympus Mons is a classic shield volcano, flattened, with a nest of six separate calderas (volcanic vents) at its summit, 37 x 50 miles (10 x 80km) across. Around its base aeolian erosion caused by wind-borne dust particles has created cliffs up to 10,000ft (3,050m) high, but at no point on the surface of the planet would it be possible to see it in its full glory – the curvature of the horizon being greater than the angle of curvature of the structure. While Olympus Mons may be the tallest, radar mapping of the planet Venus shows more than 150 shield volcanoes much flatter and broader, the largest having a diameter of 430 miles (700km).

RIGHT The cross section of a Pelean eruption, violent events characterised by streams of hot lava and pyroclastic flow which discharge large amounts of energy in a short time. The thick magma commonly forms a rounded top with steep-sided domes. Key: 1, ash plume; 2, volcanic ash rain; 3, lava dome; 4, volcanic bomb; 5, pyroclastic flow; 6, layers of lava and ash; 7, strata; 8, magma conduit; 9, magma chamber; 10, dike. *(David Baker)*

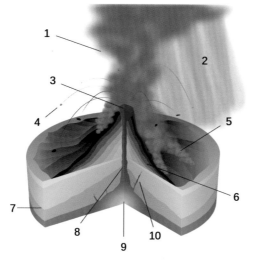

All shield volcanoes have calderas that display several episodes of eruption, which can include sporadic firework displays of fire fountains and lava sprays reaching several hundred feet into the air. Shield volcanoes are interesting to study because they incorporate a variety of features, and among all types are probably the easiest to explore and to investigate because of their more predictable activity. But not all calderas are self-evident and there is one that forms a national park in North America – Yellowstone.

Over a period of at least 19 million years, what is now known as Yellowstone has been moving gradually across a thermal plume in the mantle that has periodically erupted bringing great geographical changes to the region. Calderas that now form highly attractive parts of Yellowstone National Park testify to gigantic eruptions 2.1 million years, 1.3 million years and 640,000 years ago. This is what geologists loosely term a 'supervolcano', and the most recent analysis indicates that is has a magma chamber measuring 50 x 12 miles (80 x 20km) containing 960 cubic miles ($4,000km^3$) of mass with less than 8% filled with magma. Earlier estimates had calculated it to be only 40% the size it is now known to be, but scientists also believe that it contains far too little molten lava in the chamber to produce another eruption.

Comprising cinder, ash and lava, stratovolcanoes are most commonly found around subduction zones and are believed to be unique to Earth, although there is dispute about the classification of Zephyria Tholus on Mars. They are arguably the most dramatic of all volcanoes, having encouraged filmmakers to base dramatic stories around their explosive eruptions, Krakatoa being one. The most recent eruption occurred in August 1883, discharging 6 cubic miles (25km³) of rock in an explosion that was heard 3,000 miles (4,800km) away in Alice Springs, Australia. It killed at least 36,000 people, destroyed 165 villages and small towns and created tsunamis which went on for several hours, killing an additional unknown number. Located in the Sunda Strait between the islands of Java and Sumatra, Krakatoa is but one of more than 130 active volcanoes across Indonesia, but it was certainly the most dramatic in recorded history and perhaps the best known because of the misnamed 1969 film, *Krakatoa – East of Java*. It lies to the west!

Stratovolcanoes can be found around the continental crust margins and in arcs around Japan and the Aleutian islands. They have dramatic and explosive eruptions from the way their constituents react. The magma forms when water trapped in hydrated minerals and in porous basaltic rock from the ocean crust is released into mantle rock from the asthenosphere above the descending ocean slab. The drying process occurs at different temperatures and pressures for each mineral, and water that is liberated from the rock lowers the melting point of the overlying mantle rock. This undergoes partial melting, and because it has less density it begins to rise compared with the surrounding rock, pooling temporarily at the base of the lithosphere.

As the magma rises through the crust it incorporates silica-rich crustal rock and when it reaches the top it gathers in a magma chamber directly under the volcano. While there the lower pressure causes the water to dissolve along with other volatiles such as carbon dioxide, sulphur dioxide and chlorine, which escape from solution, building up pressure like a shaken champagne bottle. Each volcano will have a critical point at which the explosive eruption is triggered and the cork will pop, sending vast quantities of material propulsively discharged by the excess pressure.

While volcanoes can be dramatic and

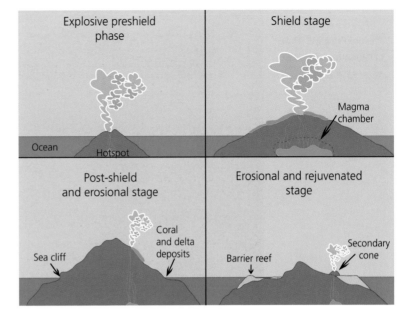

the consequences of their activity very often the stuff of Hollywood disaster movies, the consequences in our world today are very real and peculiarly apposite. Because volcanic rock is rich in silicon, which is readily bonded with oxygen, these two elements make up 75% of Earth's crust and so there is much bonding within the magma. The explosive discharge that comes from the solid compound SiO_2 is partially crystallised, thick and viscous, but the energy released due to the capping effect – which has restrained extrusion until great pressure has built up on the thick lava – propels the contents to extreme altitude, and this has a very great effect on climate as well as weather.

Another consequence of mobile plate tectonics is the earthquake, which can also be caused by general settling, unsettling or adjustment of blocks on or between continental plates. Earthquakes occur whenever there is sufficient tectonic energy to start a fracture along a fault plane, sides of a fault moving past each other and sticking due to imperfections in the face of each plane. Earthquakes at plate boundaries can generally be categorised into three particular types: normal, in areas where the crust is being extended, such as in divergent zones on the ocean floor; reverse faulting, where plates are converging and subducting; and strike-slip faults, where two steep-sided blocks are moving past each other. Sometimes movement takes place in what are described as transform boundaries, where complex

ABOVE The Hawaiian volcanoes on the seamount at the Pacific Plate evolve through a sequence identified in this graphic where the explosive preshield rises from a thermal plume (top left) and creates a shield above the magma chamber (top right). Erosion from sea and air reforms the geography (bottom left) after which reefs may form and secondary cones erupt. *(David Baker)*

interaction between different angles of slip can
cause what geologists call an oblique slip.

Reverse fault earthquakes are by far
the most violent because the magnitude
is proportional to the area of the fault that
ruptures. The longer the dimensions of length
and width of the fault, the greater the resulting
magnitude. Only the cool and brittle rocks at
the very top layers of the crust can produce
earthquakes because at a temperature of and
above 572°F (300°C) rocks become malleable
and flow rather than stick. The longest

faults that have broken in a single rupture
are examples observed in Chile, Alaska and
Sumatra, where 600-mile (1,000km) fractures
are known. Strike-slip faults are much shorter,
while normal faults are shorter still. The most
important factor is not the length of the fault but
the width. In subduction zones, for instance,
where the dip angle is usually no more than
10°, the width of the brittle upper plane can be
as much as 66 miles (100km).

Earthquakes are today measured using a
system devised in 1935 by Charles Francis
Richter and Beno Gutenberg of the California
Institute of Technology. Their Richter scale is
based on the scale of apparent magnitude
for stars, showing a magnitude 0 event as
displaying a horizontal displacement of 1.0µm
(0.00004in) recorded on a seismograph at a
distance of 62 miles (100km) from the epicentre
– the spot on the surface directly above the
point where the earthquake occurred, known as
the hypocentre. Initially it ranged through a scale
from 1 through 9 with each number raising the
magnitude of the earthquake by a factor of 30.

Using this common scale of specified
distance, the real magnitude of the earthquake
could be calculated. The unit of magnitude
0 was adopted as the base despite it being

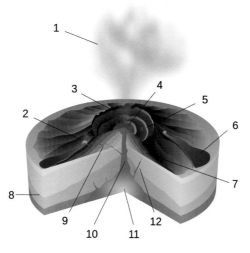

a level undetectable with the instruments of the day but it was a level they knew could be measured when instruments with that sensitivity became available. The system had a built-in margin for later, more technologically advanced, developments. In reality, in the decades since it was introduced seismology has come a long way and instruments today can measure down to tiny fractions of a magnitude.

The general scale of impact rises from measurements of less than 2.0, which is classed as a micro-quake, which is not felt or observed by normally sensitive people. Magnitude 2.0–3.9 earthquakes are classed as minor and involve no more disturbing effect than indoor objects shaking. Magnitudes 4.0–4.9 are light and involve some objects being displaced from shelving, etc. While magnitude 5.0–5.9 causes damage, 6.0–6.9 is classified as strong and will damage buildings, and will be felt hundreds of miles/ kilometres from the epicentre. Magnitude 7.0–7.9 will cause damage to most buildings, cause collapse and bring down unstressed structures, while on the scale of 8.0–8.9 structures will be destroyed and damage will occur to earthquake-resistant buildings. Magnitude 9.0 and above imply close to total destruction and permanent changes to the topography.

This original scalar classification of earthquakes and their effects has been extended at both ends of the scale and seismometers can now measure minuscule perturbations in the ground up to massive eruptions far beyond the registered levels of the original scale. In terms of the amount of energy released, a large hand grenade would induce a magnitude 0.2 reaction while a single stick of dynamite with an explosive charge of 2.4lb (1.1kg) would register 1.2 on the Richter scale. The Oklahoma City bombing of 1995

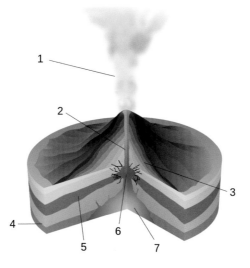

ABOVE Mauna Kea as viewed from the Mauna Loa Observatory, Hawaii. Note the shallow flank slope and curved domical cap. *(David Baker)*

LEFT A phreatic volcano is created in water where the high temperature of the magma encounters a fluid and explodes with hot steam, throwing out ash, rocks and volcanic bombs. Paradoxically, the rocks hurled from this type of volcano can be cold. Key: 1, water vapour cloud; 2, volcanic bomb; 3, magma conduit; 4, layers of lava and ash; 5, water table; 6, explosion; 7, magma chamber. *(David Baker)*

ABOVE The Thurston lava tube on Big Island, Hawaii. Lava tubes are caused by lava running through subsurface cavities or by lava excavating an underground tunnel which cools, contracts and leaves a void. *(David Baker)*

recorded 3.0, a Kent earthquake in the UK was measured at 4.3 while an earthquake in Lincolnshire in 2008 registered 5.0.

The most powerful earthquake in recorded time was registered in Valdivia, Chile, in 1960,

RIGHT A subglacial volcano occurs under a large quantity of ice, the heat from the lava melting the overlying cover to produce meltwater which hardens into pillow-shaped lava. Key: 1, water vapour cloud; 2, lake; 3, ice; 4, layers of lava and ash; 5, strata; 6, pillow lava; 7, magma conduit; 8, magma chamber; 9, dike. *(David Baker)*

recording 9.5, with the epicentre 350 miles (570km) south of Santiago. This tremor caused local tsunamis up to 82ft (25m), which reached across the Pacific Ocean and devastated Hilo, Hawaii, with waves up to 35ft (10.7m) high recorded 6,200 miles (10,000km) away. The US Geological Survey calculated a death toll of more than 3,000 people, while some sources quote in excess of 6,000 being killed. But this terrible tragedy in human cost is eclipsed by a considerable margin when calculating the seismic disturbances caused not by earthquakes but by meteorite impacts. The Chicxulub impact 66 million years ago, equivalent to the release of 100,000 gigatonnes of TNT, would have recorded a magnitude of 13.0 on the Richter scale.

Most earthquakes are not of high magnitude. Each year Earth experiences about 500,000 earthquakes of which only one fifth can be felt. Certain places around the world are notorious for frequent earthquakes, namely Mexico, Guatemala, Chile, Peru, Iran, Pakistan, Turkey, New Zealand, Greece, India, Italy, Nepal and Japan. But there is a proportionality ratio between magnitude and frequency. For

instance, about ten times as many magnitude 4 earthquakes occur as magnitude 5. The continuous monitoring of earthquakes through seismic stations has grown enormously over the decades, and knowledge through monitoring, advanced technological tools and better scientific understanding of these potentially devastating events has grown accordingly.

While earthquakes can occur anywhere, any time, there are specific zones and areas of the planet where their occurrence is a forgone conclusion. One such is the so-called Ring of Fire, where earthquakes and volcanoes originate from a common cause and are among some of the more violent events. The Ring of Fire is a popularised name for the 24,900-mile (40,000km) circum-Pacific seismic belt where 90% of the world's earthquakes, and 81% of the most severe, occur. It extends from the Nazca Plate and the Cocos Plate subducting beneath the South American Plate moving westward, up around the western side of North America, which is subducting the Juan de Fuca Plate, taking in the Pacific Plate moving north-west under the Aleutian Islands. From there, rotating anticlockwise around the map, the Ring includes the Pacific Plate

ABOVE **Lava domes on Mount St Helens, an active stratovolcano in Washington, most notorious for its dramatic eruption in 1980. It developed into the most destructive in US history, killing 57 people and causing widespread damage.** *(David Baker)*

subducting under the Kamchatka Peninsula arcs and on past Japan. South of Japan the Ring gets complicated due to the profusion of small plates colliding with the Pacific Plate in the Mariana

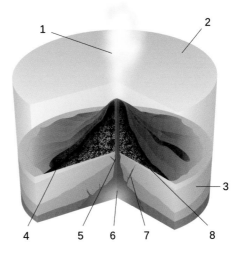

LEFT **Unlike a subglacial volcano, a submarine volcano erupts where there is almost limitless water for cooling the lava. They usually occur at mid-ocean ridges – about 75% of all volcanoes take this form. Shallow eruptions can cause small islands to form, as with Surtsey off Iceland. Key: 1, water vapour cloud; 2, water; 3, stratum; 4, lava flow; 5, magma conduit; 6, magma chamber; 7, dike; 8, pillow lava.** *(David Baker)*

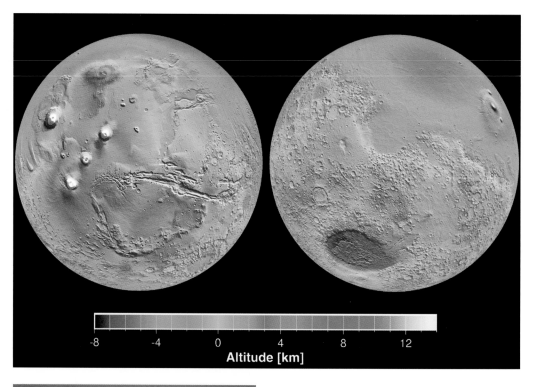

RIGHT Volcanoes on Mars can build to greater size than they do on Earth because there is no mobility in crustal layers. Mantle plumes punch through the same area of the crust over very long periods, building up enormous domical upwellings and massive volcanoes. *(NASA)*

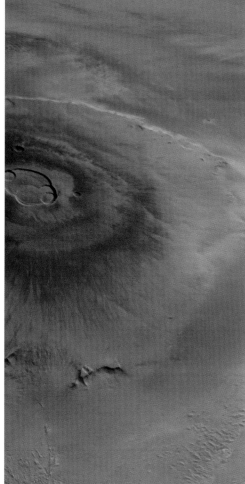

RIGHT The caldera on Olympus Mons, the largest volcano on Mars and in the solar system, three times the height of Mount Everest above sea level, encompasses several episodes of eruption. *(NASA)*

Islands, the Philippines, Bougainville, Tonga and New Zealand.

This south-western area of the Ring of Fire is particularly interesting because it incorporates the Mariana Trench, the deepest part of the oceans. Stretching for 1,580 miles (2,550km), it runs in a hooked arc connecting the southern Japanese trench with a position just north of Indonesia and is named after the Mariana Islands and in honour of the Queen of Austria, widow of Philip IV of Spain (1605–65). Geologically it is part of the Izu-Bonin-Mariana system forming the boundary between two plates, with the western edge of the Pacific Plate subducting under the Mariana Plate, which is smaller and is positioned to the west.

The trench has an average width of only 43 miles (69km) and the greatest known depth at a single point of measurement was recorded as 36,170ft (11,025m), while the maximum consistently recorded depth, at a place known as Challenger Deep, was only a little shallower. The pressure of the water at this depth of almost seven miles is 1,086 bar (15,750psi/108,596kPa) but the density is increased by almost 5% (34.475kPa), and the temperature is a chilling 33.8–39.2°F (1–4°C). Several successful attempts have been made to reach the bottom, the most notable using the bathyscaphe *Trieste*, which

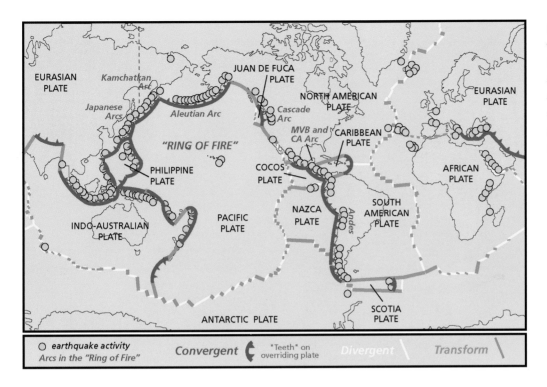

Map labels:

EURASIAN PLATE · Kamchatkan Arc · JUAN DE FUCA PLATE · NORTH AMERICAN PLATE · EURASIAN PLATE · Japanese Arcs · Aleutian Arc · Cascade Arc · MVB and CA Arc · CARIBBEAN PLATE · AFRICAN PLATE · "RING OF FIRE" · PHILIPPINE PLATE · COCOS PLATE · INDO-AUSTRALIAN PLATE · PACIFIC PLATE · NAZCA PLATE · SOUTH AMERICAN PLATE · Andes · ANTARCTIC PLATE · SCOTIA PLATE

Legend:

○ earthquake activity — Arcs in the "Ring of Fire" · **Convergent** "Teeth" on overriding plate · *Divergent* · *Transform*

completed the first descent in January 1960. The most recent of the four descents was completed by film producer James Cameron in March 2012 when he reached the bottom in the submersible *Deepsea Challenger*.

This generalised view of the Phanerozoic, the last 541 million years, has taken a broad approach to describing the events that bring us to the present day in the geological activity of our planet. But this has so far been a one-sided story, for Earth is more than the summed products of its geology and its plate tectonics. Earth is unique in our consciousness because it is the only planet we know for sure which has advanced forms of life. There were many stages to the development of life across this post-Precambrian period and those changes will be observed in Chapter 7, where the changing atmosphere was driven by the evolution of life.

The changing Earth is a dynamic and evolving engine, fed by fuel from within the mantle, transforming the surface through modification to the atmosphere and diversifying life in its migration from water to land. Life is found everywhere on Earth and the prospect of life on this planet being an analogue for life elsewhere has encouraged a broad study on the possibility for what is now known as exobiology. But much of what happens in the development of life happened in the seas and oceans of the world, which is where its weather is largely formed, and the evolution of the seas and oceans is our next journey through the history of Earth.

Summary

- The early crust was very different to the later crust, rearranged by the Late Heavy Bombardment.
- The response of rocks to temperature is determined by the amount of water to which they are exposed.
- Subduction is key to Earth's ocean crustal mobility back into the mantle.
- The Wilson Cycle lays down the mechanisms for ocean crust production and subduction.
- A series of supercontinents has formed to significantly modify Earth's climate.
- Supercontinents form on average every 400–450 million years.
- Earth has frequently been an icehouse planet with almost total global glaciation.
- Major meteorite impacts and mantle plumes periodically remodel large surface areas.
- The volcanic record verifies plate tectonics and a giant volcano on Mars demonstrates the fact.

Chapter Five

The oceans

We have seen from Chapters 3 and 4 that water has covered most of Earth throughout its history, back to the post-accretion phase more than 4 billion years ago and while continental materials were still forming, long before the great bombardment transformed a settling planet into a very new world. Water has been fundamental to Earth's surface condition and it has changed the way rocks behave under different temperatures and pressures.

OPPOSITE A picturesque gully in the Mariana Arc region where red and green algae and soft pink corals with white stalks provide scenic routes for tropical fish. *(NOAA)*

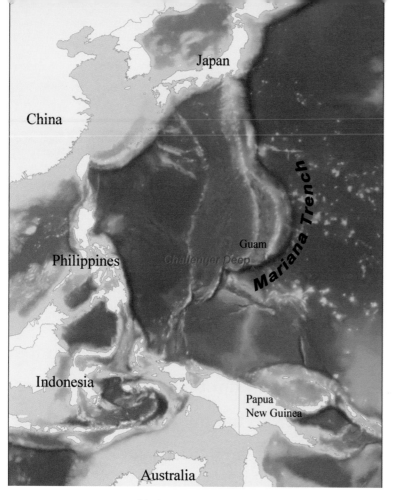

Japan

China

Philippines

Guam

Challenger Deep

Mariana Trench

Indonesia

Papua
New Guinea

Australia

ABOVE Some of the deepest and oldest parts of the ocean crust are located adjacent to subduction zones, the Mariana Trench system being the deepest on Earth.
(NOAA)

Hydrous rocks respond differently to anhydrous rocks and water itself helps energise the tectonic system. Water is crucial to life, and of that there has been great abundance in the world's seas and oceans. It has been an empowering force for climate and weather, be it in liquid, gaseous or solid form. In fact, water is the only substance that can occur naturally and unmodified in those three states.

In physical terms, water is a fluid and so is air – one in liquid form, the other as a gas. Both water and air are subject to the same principles of fluid dynamics and because of that an understanding of one aids in interpreting the activities of the other. Boat designers make good plane-makers, and this is why, in the early days of aviation, many winged machines that took to the air did so from water – the latest application of mankind's urgent desire to explore and to climb the next hill by combining two fluids that have controlled extreme environments in the story of Earth.

Water is a liquid across a wide range of temperature, buffered between the solid phase at one end and evaporation at the other. These two extremes, turning water into a solid and

a gas, frame the temperature scale in Celsius, 0°C being the freezing point and 100°C being the boiling point. Water flows from springs, streams, rivers, seas and oceans and covers 70.9% of Earth's surface – almost completely covering the ocean crust, shorelines being only a little further inland than the edges of the continental crust.

Sea levels rise and fall according to the quantity of water locked up in the solid phase, and that has gone up and down between dramatic extremes for billions of years. Our shorelines today are much lower than at many times in Earth's history, and much higher than at others. Cyclical movement is governed by climate and by crustal motion inducing changes to the balance between liquid and solid phases, but also on a daily basis by the movement of the Moon.

Of all the water on Earth, 96.5% is in the seas and oceans with 1.7% in groundwater and a further 1.7% in the ice caps at the North and South Poles and on Greenland and in glaciers across the world. Of Earth's total water content only 2.5% is freshwater, of which only 0.3% is in the ice caps, glaciers, rivers, the lakes and the atmosphere, with a factional 0.003% of all freshwater in organisms, including humans. Water is essential for all living things but increasingly it is used to support forced agriculture by humans, 70% of all freshwater going into growing food and for manufacturing food products.

Water has been used as a tool by humans for millennia, water-powered wheels turning millstones and grinding tools, and for producing energy, from the large-scale use of high-pressure water to drive turbines producing electrical power, to the small-scale use of water to produce electricity. Through the process of electrolysis an electric current can be used to split water into its constituent molecules, hydrogen and oxygen; and through reverse electrolysis, combining hydrogen and oxygen over a catalyst, it can produce an electric current – the 'fuel cell', used in spacecraft since 1965 and motor vehicles more recently.

The physical properties of water are crucial to an understanding of its role in Earth, as it has the property to dissolve more substance than any other liquid and has the highest surface tension, as well as the highest thermal capacity for storage, thermal conductivity and heat of vaporisation, of any common substance. It has characteristics used as a yardstick for measuring density and it has high sound transmission qualities compared with other liquids. It is virtually incompressible and has unusually high boiling and melt points. Because of these properties it is an effective coolant, not least for water-cooled reciprocating engines, which some might say includes Earth.

Water consists of two hydrogen atoms and one oxygen atom covalently bonded together, and because of this it has properties that allow high absorption levels by the hydrogen-oxygen (OH) bonds bent in a nonlinear (elongated) geometry. The oxygen atom carries a slightly higher electron negativity than the two hydrogen atoms, which are slightly positive, and this conveys an electrical dipole moment on the water molecule. The bond dipole moments do not cancel each other out and this sets up a molecular dipole with the negative pole at the oxygen and its positive pole at the mid-point between the two hydrogen atoms.

The density of liquid water is 1g/cm³, which is also the magnitude of density for flotation. Thus we can say that, were there to be an ocean large enough to contain the planet Saturn, with a mean density of 0.687g/cm³, it would float – the only planet in the solar system to do so! Water is also a product not only of reverse electrolysis to produce

energy, but of the burning of hydrogen. As an oxidising propellant to hydrogen fuel in the case of a rocket engine, as used to power liquid hydrogen/liquid oxygen rockets today, it produces steam from the combination of these two atoms in the process of combustion.

A property of seawater in the world's seas and oceans is its salinity, about 3.5%, which is 35g/L or 599mM (1 millimolar = 10^0mol/m³,

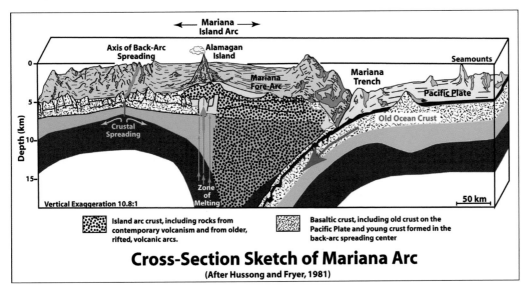

Cross-Section Sketch of Mariana Arc
(After Hussong and Fryer, 1981)

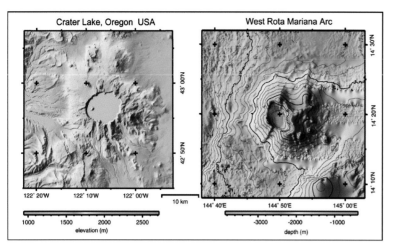

Crater Lake, Oregon USA

West Rota Mariana Arc

ABOVE A comparison of similar calderas for Crater Lake in Oregon, USA, and West Rota Mariana volcano where the former is shown in elevation above sea level, the latter in depth below the surface. *(NOAA)*

where a molar is a measure of the solute in a litre of solution and measured as mol/m³). In other words, each kilogram of seawater has 35gm of dissolved salts consisting of sodium (Na⁺) and chloride (Cl⁻) ions. The average density of seawater is 1.025g/ml, while freshwater has an average density of 1.00g/ml. At this density level of 35g/kg the thermal conductivity of seawater is 0.6W/mK at 25°C (77°F), which will increase as the temperature rises and decrease as the salinity goes up.

Saline values around the world vary but are usually in the range 3.1–3.5% while the density is 1,020–1,029kg/m³, varying according to the temperature and pressure. At high pressure, deep in the ocean, this level can reach 1,050g/m³. Famous for its attraction for bathers, the Dead Sea is the most saline open tract of water in the world, made so because of the contained circulation, high levels of evaporation and low precipitation. Saline levels also determine what is freshwater, because that too contains low levels of salt but at insufficient quantities to be limiting in its use. Freshwater is defined as having less than 1,000mg/l of dissolved salts, but it is still there.

The salinity of the world's seas and oceans has been stable for several billions of years and was initially established when the primordial ocean began to leach out sodium from the seabed. Chloride, the other dominant ion in salt, was introduced from the outgassing of hydrochloric acid from hydrothermal vents in the sea floor and from volcanoes. This interaction between the tectonic activity of the asthenosphere and the lithosphere has

RIGHT The oceans and seas provide the energy that drives much of the planet's weather. The water cycle characterises that activity, where evaporation from the oceans leaves salt behind and creates freshwater rain to precipitate over land and flow back down to the ocean. *(Met Office)*

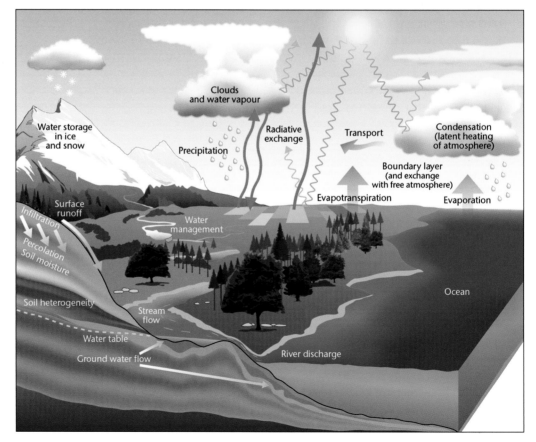

established a balance between supply and deposition, sodium and chloride sinks being responsible for removing as much as is laid down. Chemical reactions also limit the quantity, including interactions with ocean basalt.

For humans and other animals, saline water is harmful and cannot be drunk as ordinary freshwater can. The saline content of seawater is far too high for the kidneys to process if taken undiluted, but in extreme situations a diluted intake of one cup of seawater to two cups of freshwater will extend the use of the latter and delay the onset of dehydration without permanent damage to the body.

In this chapter we are dealing with seas and oceans but in the next we will cover the atmosphere, which is where the freshwater cycle begins. But to second-guess an urgent question that could arise here, the water we drink and all the freshwater in the world, while coming from evaporation off seas and oceans, leaves the salt behind and rises without the sodium and chlorine molecules attached. When it precipitates out as rain, on mountains, into rivers and back on the sea itself, it is fresh, as we say. Which is the reason why rain which comes from the seas and oceans is good to drink.

Water world

The 70.9% of Earth covered with water occupies a surface area of 139.7 million square miles (361.9km² million), which supports a volume of 323.3 million cubic miles (1.335km³ billion). Yet the total hydrosphere accounts for a mere 0.023% of Earth's mass and the entire water content of the planet would occupy a cube 684 miles (1,101km) on each side. Despite this, about 66% of the surface of Earth is covered by deep ocean zones, defined as anything below 660ft (200m) in depth, with 50% of all the world's oceans deeper than 9,800ft (3,000m). As related at the end of the last chapter, the lowest part of ocean floors is found at the subduction zones, the Mariana Trench taking the record.

The world's oceans are classified as comprising the Pacific, the Atlantic, the Indian, the Southern and the Arctic oceans. The Arctic Ocean was added in the year 2000, and is defined as all the waters surrounding the Arctic

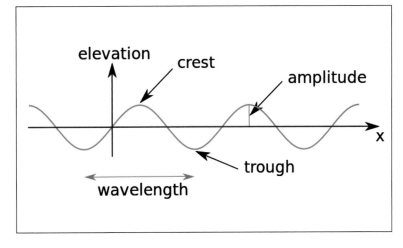

ABOVE Sea state is measured by wavelength between troughs and elevation in height, the amplitude of the wave being the mean height above sea level. *(David Baker)*

BELOW Seawater is approximately 3.5% saline. This salinity results from a range ofcomponents, among which chloride is dominant, accounting for more than half. *(David Baker)*

BOTTOM A given quantity of liquid water takes up less volume than the same volume as ice because the molecules are arranged in a distributive bond. *(David Baker)*

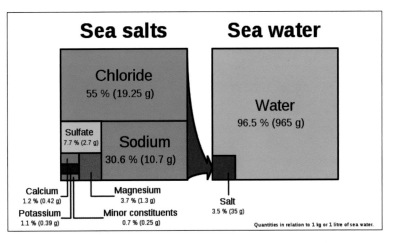

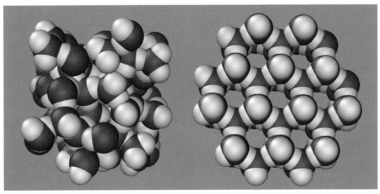

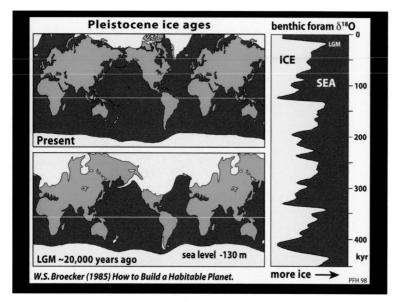

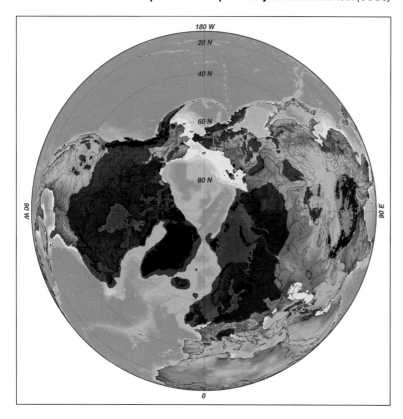

ABOVE Ice ages frequently change dramatically the level of the oceans and open seas, exposing expansive offshore areas which had supported evolving marine life forms for millions of years, sea levels in some instances falling several hundred feet (tens of metres). This graphic compares sea levels during the Last Glacial Maximum (LGM) with those today, still rising as a consequence of the ice age. *(USGS)*

BELOW A graphic representation of the levels of ice cover during the Last Glacial Maximum (in shadow), with Arctic areas covered today (in black), reveals a wide expanse of land previously covered with ice. *(USGS)*

down to latitude 60°N. The divisions between the five oceans are entirely arbitrary but the biggest by area and volume is the Pacific Ocean, which accounts for 43.5% of them all, followed by the Atlantic Ocean at 28%, the Indian Ocean at 19.38%, the Southern Ocean with 5.36% and the Arctic Ocean with 3.69%. Of course, all these oceans are connected so the divisions are largely for classification, as they are all open. Seas, for their part, are largely defined as tracts of water completely or partially closed but mostly with an open access to ocean.

The oceans are divided into zones according to depth and light absorption factor. The pelagic zone includes all surface areas connected by low sea-level lines, and from there down the zones are defined by depth and temperature. The topmost zone is known as the mesopelagic, defined as the depth at which the temperature is at 54°F (12°C), which is usually around 2,300–3,300ft (700–1,000m). This is known as the first thermocline, a term used to describe a place where the temperature makes a significant and sudden change. Below that is the bathypelagic, which is on average down to a depth of 6,600–13,100ft (2,000–4,000m). In this zone the temperature is typically 39–50°F (4–10°C). The next deepest zone is known as the abyssopelagic, which extends to a depth of 20,000ft (6,000m), and the last zone is known as the hadalpelagic, down to the deepest levels of the ocean trenches up to 36,000ft (11,000m) below the surface.

The vast expanse of underwater crust that lies between the mid-ocean ridges and the subduction zones at the edge of the continental crust is known as the abyssal plain. It alone accounts for 50% of Earth's surface and it is here that oceanographers have mapped the vast expanses of gently undulating terrain where large quantities of sedimentary material stripped from sea floor vents and continental shelves deposit silt and clays, where reservoirs of biodiversity reign supreme. Concentrations of metallic nodules, manganese, nickel, cobalt and copper litter the deposits, while the organic material is regulated by the differing quantities that make their way to the sea floor.

There are other systems scientists use to divide up the layers in the ocean, notably by density, these being the surface zone, the

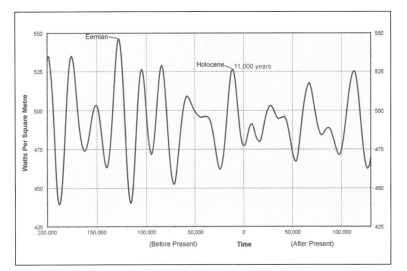

RIGHT Insolation levels showing the amount of sunlight falling on Earth and its oceans as measured in watts per square metre over the last 200,000 years and into the future *(David Baker)*

CENTRE Rates of sea level rise and fall coincident with the solar cycles from the early part of the 20th century to 2000. *(David Baker)*

BOTTOM Past and future glacial maxima based on the cyclical trends connected to the obliquity of Earth's orbit and the precession of the equinoxes. During these cycles the oceans go through dramatic transformation which in turn greatly influences the weather. *(David Baker)*

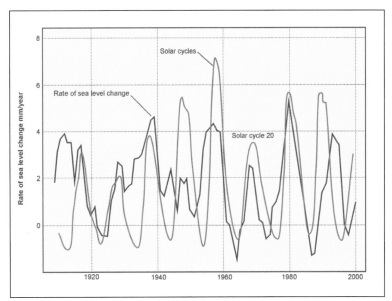

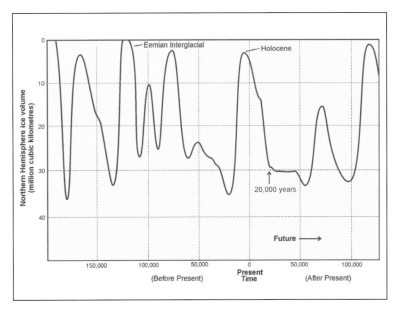

pynocline and the deep zone. The topmost, or surface, zone is a mixed interplay of surface winds, upper currents and thermal turbulence due to atmospheric conditions, but the overall temperature and salinity levels remain comparatively stable. This is where the ocean meets the atmosphere and can extend anywhere from 500ft (152m) to 3,300ft (1,000m), but can vary between northern latitudes and the tropics. This zone is a mere 2% of the ocean. The transition to colder and denser conditions marks the boundary between the surface zone and the pynocline.

It is here that the density increases significantly, while the temperature decreases with depth and is the dividing line between the upper and lower density regions. This part represents approximately 18% of the ocean, fluctuating according to prevalent conditions. It is turbulence from wind and wave at the top of the surface that is transferred down as heat, and the temperature at the surface/pynocline boundary can be 36°F (20°C). This thermocline is steep and meets the underlying, much colder temperatures at the deep zone.

Excepting the mesopelagic, the deep zone is devoid of light, and plants are unable to exist because there is an absence of opportunity for photosynthesis. As pressure increases by 1atm for each 33ft (10m) of depth, marine organisms at this extreme depth fight crushing forces unimagined by life on the surface. This deep-sea zone is highly stable – in temperature and pressure, of course, but also in salinity levels. It is

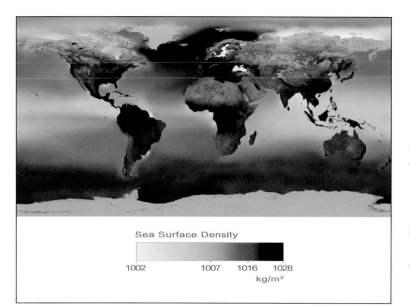

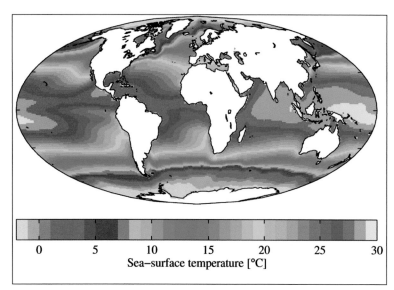

Sea−surface temperature [°C]

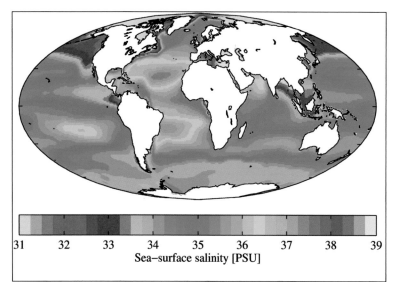

Sea−surface salinity [PSU]

a still, virtually immobile world where simple and highly inventive applications of evolution produce life forms unimagined by scientists until exploration applied eyes and cameras to these depths.

Variations in the division and zoning can eliminate entire sections of these stratigraphic regions, particularly in northern waters where there is little sunlight to warm the water and where there is little vertical difference in the temperature. In such waters, below the thermocline the water is exceptionally cold, from 30° to 37°F (-1° to 3°C). For this reason the average temperature of the world's oceans is 39°F (3.9°C). Differences in temperature and density are the primary reasons why the oceans are turbulent and it is from this that ocean circulation is driven. Just as the conveyor belts of ascending magma from Earth's mantle drive plate tectonics so do the global density gradients in the world's oceans drive surface heat and currents.

The large-scale movement of water in the oceans by this mechanism is known as thermohaline circulation, a word derived from 'thermos' and 'haline', because temperature and the saline content drives density, and that more than anything else creates the large-scale transfer of water around the world. While wind circulation causes the apparent motion of surface water, the real driver is the changes in density within the ocean itself. As we have seen, the density varies across the globe and because of that the ocean current gains traction from this motivating effect.

Sinks and basins

As the ocean warms it expands and becomes less dense, and changes also occur with varying levels of salinity. This causes flows to occur and sets up deep-water currents. The formation of deep-water masses in the North Atlantic and in the Southern Ocean due to thermoclines also stratify the oceans, the lighter water mass floating across the top of the heavier water. Dense water sinks back down at the polar ocean basins and is therefore offset by water rising in other places, but the rates of rise and fall differ, the heavier water sinking comparatively quickly and warmer water rising more slowly. Vertical columns of rising water move at about 0.5in (1cm) per day, but here the upward motion can be arrested by capping due to temperature inversions on the surface. The two great basins for deep-water masses are the North Atlantic and the Southern Ocean.

Although all the oceans are interconnected, land masses influence the way the great reservoirs of dense water develop, and that is due to evaporative cooling. This occurs when surface water affected by wind moving across the uppermost layer causes evaporation, leaving the heavier salts behind and increasing the saline level, and the density, as a result. Evaporation of water molecules in the North Sea (or Norwegian Sea) between the British Isles and Norway increases the density due to the increase in salinity, and this water sinks as it moves into the Atlantic through the Greenland–Iceland–Scotland ridge across the Faroe Bank channel via underwater sills.

This southward motion is lethargic but remorseless as the weather in the North Sea keeps active a system that seeds the great abyssal plains in the Atlantic Ocean. But saline-rich water from the Arctic Ocean is prevented from flowing south into the Pacific Ocean by the Bering Strait, which is too narrow for the denser mass to move through. This creates a circulatory flow that benefits the movement of deep water across to the North Sea and down that route into the Atlantic.

In the southern hemisphere the situation is the reverse. Whereas the North Pole has no land mass and any surface is formed from ice, the South Pole is located on the continental crust of Antarctica, and the circulatory motion of the Southern Ocean is influenced by strong winds from the continent itself on to the ice sheets and off to the open ocean. There it chills the saline seawater which starts to freeze at 28.7°F (-1.8°C), and because seawater does not have the density base of freshwater at 39.2°F (4°C) it continues to densify. As the ice begins to reform from the colder air and seawater temperature it produces saline-free icebergs and shelves, leaving saltier brine behind in the free-running water as it solidifies.

Thus formed, the dense water sinks as it flows north and east but its density is so great that it flows beneath the North Atlantic Deep Water (NADW), which has continued to move south from its abyssal sink. The southward

BELOW Another way of measuring salinity is in units of parts per thousand, adopted here for this graphic showing relative proportions of salinity for water from different sources. *(David Baker)*

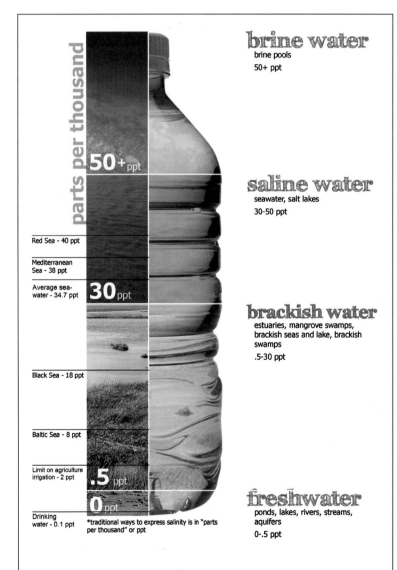

parts per thousand

brine water
brine pools
50+ ppt

50+ ppt

saline water
seawater, salt lakes
30-50 ppt

Red Sea - 40 ppt

Mediterranean Sea - 38 ppt

Average sea-water - 34.7 ppt

30 ppt

brackish water
estuaries, mangrove swamps, brackish seas and lake, brackish swamps
.5-30 ppt

Black Sea - 18 ppt

Baltic Sea - 8 ppt

Limit on agriculture irrigation - 2 ppt

.5 ppt

0 ppt

Drinking water - 0.1 ppt

*traditional ways to express salinity is in "parts per thousand" or ppt

freshwater
ponds, lakes, rivers, streams, aquifers
0-.5 ppt

RIGHT The ocean depths are mapped according to conditions in the water regarding depth and pressure and temperature differences, from the epipelagic to the lowest level at the hadalpelagic. *(NOAA)*

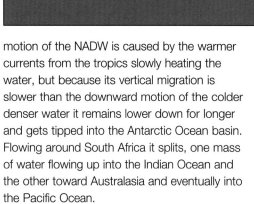

BELOW The terminology displayed here defines various subsurface land forms down to the ocean floor. Oceanographers use continental and water definitions for plotting currents and charting the mass movement of water from surface levels down to the depths. *(NOAA)*

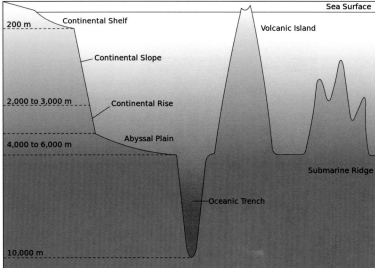

motion of the NADW is caused by the warmer currents from the tropics slowly heating the water, but because its vertical migration is slower than the downward motion of the colder denser water it remains lower down for longer and gets tipped into the Antarctic Ocean basin. Flowing around South Africa it splits, one mass of water flowing up into the Indian Ocean and the other toward Australasia and eventually into the Pacific Ocean.

Another fascinating aspect of this is the varying levels of the sea, which is higher in some parts of the world than others. Sea level is generally taken to be a constant elevation at any point around Earth – water always finds a common level, it is said. But that is not true on the large scale of seas and oceans. We saw in Chapter 4 how the inconsistencies in the bulk structure of Earth produce areas of denser mass than at others, and this has an effect on the gravitational attraction at various locations around the planet. This also influences the flow of water between areas with different gravity levels, as well as on the movement of seawater at the surface.

Sea-level measurements are taken as the mean of the average around the globe and that has been rising over the last several thousand years due to the melting glaciers at the end of the last ice age. In that time sea levels have risen about 394ft (120m), stabilising out around 2,000–3,000 years ago. After consistently remaining at that level, they have been rising during the 20th century at the rate of approximately 0.1in (1.8mm) per year, or 10in (255cm) over a century.

This is primarily due to three factors: the rise

in sea level temperature causing an inflation in the upper 1,640ft (500m); water sources on land flowing out from melted glaciers; and extraction of groundwater for agricultural activity and from hydrocarbon fuel extraction such as shale mining. About 65% of the sea-level rise is due to expansion of the oceans with about 28% coming from outflow. A large number of minor contributions come from displacement of continental materials, etc.

Variations in the height of the sea can be measured down to very precise values, data being collected largely by satellites which use laser altimetry to record the time taken for a laser beam to reach the surface and be reflected back, the time difference allowing the exact distance to be calculated. This sea-surface topography maps sea-surface height relative to Earth's geoid, which is calculated as an equal gravitational potential and records the surface as though it were not moving. These topographic variations are caused by waves, tides, currents and atmospheric pressure, and measurements define the flow of currents with their troughs and peaks. Direct measurement helps to quantify the magnitude of the sea state while averaged data sets of multiple data-takes allow the mean average to be calculated from which the height of the surface from the mean radius of Earth can be derived.

The sea state is a defined condition that takes into account wind, waves and swell. Waves are defined by local wind conditions while the swell is a series of waves caused by distant weather systems, which are technically referred to as mechanical waves. This data is vital for providing information to mariners regarding the state of the sea as measured against an internationally accepted set of codes rating all conditions from 0 to 9. A zero sea state is one in which the sea has no wave height and is glassy calm. Intermediate levels of sea state go from wave heights of up to 0.33ft (0.1m), referred to as 'calm' for state 1, to waves over 46ft (14m) in sea state 9, formally classified as 'phenomenal', which is quite an apt description to which many sailors would testify!

Understanding the oceans of the world has been a hard-fought battle over access,

data gathering and dynamic interpretation of observations. Until the mid-1960s there were very few opportunities to observe the large-scale effects of complex swell patterns over vast areas of the ocean. Mariners recorded observations that appeared to defy the logic of marine science as understood at the time, but observations from space that began during the 1960s brought confirmation of large-scale events unobserved on a local or regional basis.

Some of these reported sightings were difficult for marine scientists and oceanographers to understand and included visual reports of extremely high wave forms in a linear string stretching over several hundred miles across the surface, moving at great speed. Photographs of such phenomena appeared to show walls of water several tens of yards high moving fast across fronts up to 400 miles (644km) long, appearing to rise up, travel great distances across the Pacific Ocean, and subside again before reaching land.

Such phenomena had been reported by ships encountering flounder conditions in other oceans. The Cunard ocean liner *Queen Elizabeth 2* encountered such a freak wave when a junior office noted on his ship's radar a blip that appeared to be a malfunctioning screen. Raised from his sleep the captain rushed to the bridge only to realise that a 100ft (33m) wall of water was only minutes away from striking the liner amidships, with the inevitable

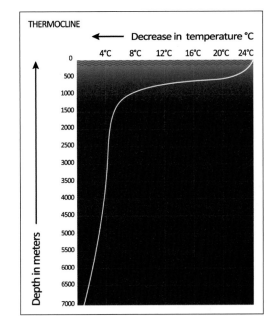

LEFT The thermocline is measured as the change in temperature with depth, a change that initially occurs rapidly. Below around 1,100–1,500m the temperature falls only slowly, however. *(NOAA)*

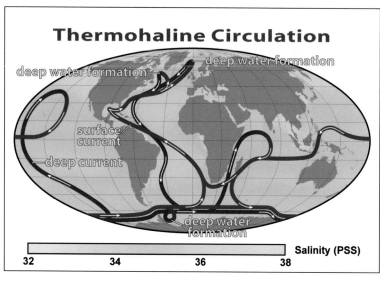

Thermohaline Circulation

deep water formation

deep water formation

surface current

deep current

deep water formation

Salinity (PSS)

32   34   36   38

ABOVE The thermocline recirculation with salinity levels and ocean currents. *(NOAA)*

consequence of rolling it completely over. Acting with judicious speed he turned the ship to face the wave which broke across the bow, tearing away one of the flying bridges that project from either side of the wheelhouse and buckling the ship by about 1in (0.4cm), which damage remained with it until retirement. The slight misalignment in a bulkhead structure was just noticeable, if you knew where to look.

Cycles and gyrations

Because Earth is spinning, the motion of the ocean currents is influenced by Coriolis forces, named after Gaspard-Gustave Coriolis

BELOW Deep sea vents produce chemical cocktails which feed the ocean floor from the hot magma through precipitation chimneys, black smokers and other vents. Particulate and gas products move into the ocean and sink to the floor, including trace elements and REE (Rare Earth Elements). *(NOAA)*

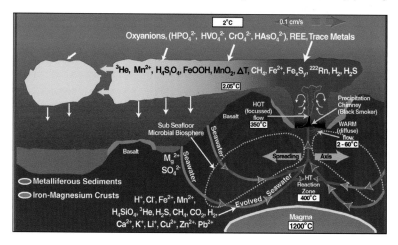

after he published a paper in 1835 exploring the energy in a rotating waterwheel. The effect is the deflection of moving objects relative to a rotating frame of reference and it has been more popularly associated with weather systems. Newtonian dynamics is referenced to a static frame within which the movement, action and reaction take place. When the frame itself is in rotation the Coriolis force comes in to play. This is defined as a force proportional to the rotation rate, while the centrifugal force is proportional to its square. The force acts outwards in a radial direction perpendicular to the axis of the rotating frame, while the velocity (speed with direction) of the mass within that frame is proportional to the object's speed. The magnitude of the centrifugal force is proportional to the distance of the body from the rotational axis. With these corrective factors Newtonian law can be applied to a moving reference frame with equal validity.

Earth with its oceans and its atmosphere is itself a moving frame of reference with an equatorial rotational velocity of 1,526ft/sec (1,040.5mph/1,674.4km/h), insufficient for the effect of the Coriolis force to be noticeable on the small scale but detectable in large circulatory systems such as water and air. The force will exert an influence for moving objects to be displaced to the right in the northern hemisphere and to the left in the southern hemisphere, greater at the poles and less at the equator. This is because the changes in latitude are much smaller toward the poles and because of this overall effect there would be a tendency for ocean currents to move in those respective directions. While there is a natural tendency for ocean water to move around according to density value at various levels, the actual motion is affected by these Coriolis forces.

Coriolis forces have been held responsible for observed differences in the way water flows down a sink drainage hole, many people being convinced that it flows one way into the drain in the northern hemisphere and the opposite way in the southern hemisphere. It is true to claim that there is, theoretically, a difference that would produce that effect. However, it is only within the purest of bowl shapes, free of minuscule surface blemishes and undulations, that an observable effect will be produced outside

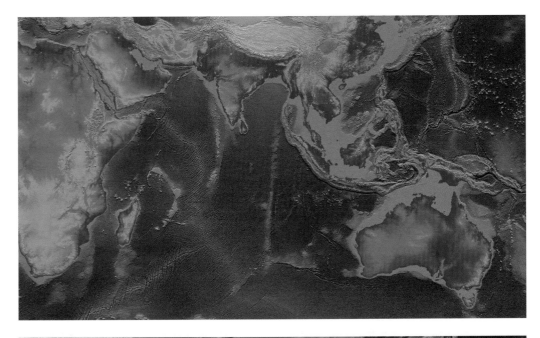

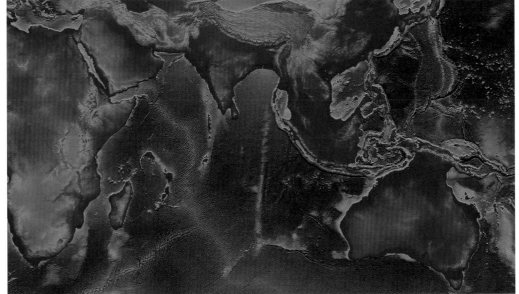

laboratory conditions. Water rotation is not detectable under normal household conditions, the scale is just too small. Once initiated, the vortex effect set up by conditions at the rim of the disc will cause the water to rotate faster as its radius diminishes, all based on our old friend the conservation of angular momentum.

This is the same physical law that dictates the way the solar system formed (see Chapter 2), but the Coriolis force is much more readily observed in the cyclonic motion of weather systems, as we will see in the next chapter. Unique to the ocean, however, is the large-scale circulatory system known as the gyre, a term more usually attached to meteorological conditions but one which defines the planetary vortices in five locations in the Indian Ocean, North and South Pacific Ocean and the North and South Atlantic Ocean.

The focus of any gyre, in air or water, is the circulation around a high-pressure zone that rotates clockwise in the northern hemisphere and anticlockwise in the southern hemisphere. It takes place in the upper 3,300–6,600ft (1,000–2,000m) of the surface and is intensified in a circulatory motion toward the pole. Each ocean gyre has a boundary current that helps condition oceanic and atmospheric events; in the case

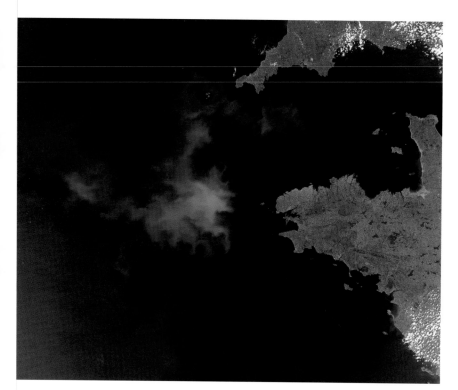

It has been calculated with due verification that every square mile of ocean surface across Earth accommodates an average of almost 50,000 tiny plastic fragments which are separated by photodegradation into small pieces which never completely break down the polymer from which they are made. These materials include low-density polythene bags, bottle caps made from polypropylene, plastic water bottles made from polyethylene terephthalate, and expanded styrofoam packaging materials. When the plastics break down into digestible sizes they are eaten by fish and other marine life that will eventually enter the human food chain. Thus, a percentage of the average food consumption by people around the world will have originated as a plastic product discarded probably decades earlier.

Only by satellite observation, direct sampling and the known degradation levels of various artificial materials in common use is it possible to determine that the dumping ground for this waste occupies approximately 10% of the Pacific Ocean. Manufacturing and the processing of consumer goods on an increasing replacement frequency is filling up the world's oceans, and that is having a disturbing effect on the way biodiversity works in the seas and oceans of the world. A potentially more serious consequence is the accretion effect at a molecular level, where chemicals that are also dumped in the seas bind with these non-degradable fragments to enter the food chain and eventually into humans.

Chemicals at greatest concern include Aldrin, Dieldrin, Endrin, Heptachlor, Mirex and Toxaphene, used as insecticides. Chlordane is another, used as a pesticide, and dioxins – a group of toxic chemicals generated as by-products from melting, smelting and paper bleaching – are also highly dangerous and mix with the polymers circulating intact and waiting for binding to these harmful chemical products. Used as a solvent, Furans is found in waste found in the Pacific gyre, together with Hexachlorobenzene, produced as a fungicide. Polychlorinated biphenyl products are used as coolants and lubricants, and these can have a disproportionate and negative impact on marine life.

Unfortunately, the concentration on a few politically-selected target pollutants, such as

years. Overall the sediment layer varies in thickness between 3.3–230ft (1–70m) above the ocean crust, its thickest accumulation being at the periphery of the gyre. Sadly, these ocean gyres are not devoid of refuse and debris deliberately dumped into coastal waters. Unfortunately the explosion of human populations in the last 75 years is having a devastating effect on the world's oceans.

Every year, of the 10 million tonnes of plastics dumped into the world's oceans only 7 million tonnes sink out of sight, where they do enormous environmental harm to marine life from just below the surface right down to the sea floor. Three million tonnes are added each year to the North Pacific Gyre, which receives rubbish from the east coast of the Asian continental land masses within about a year of it being discharged into the ocean. Plastic waste from the west coast of North America takes five years to reach the gyre. About 90% of all the waste thrown into the oceans is plastic and this is probably the most harmful, least biodegradable, product dumped annually. But these are generally not large pieces readily identifiable as waste from human manufacturing processes. Most of them would not be noticed from the side of a passing ship.

chlorofluorocarbons (CFCs), and the expanding industrialisation of most of the world's urban communities is forcing demand for exotic materials and chemical combinations which are damaging the health of the world's oceans. Moreover, in aligning low CO_2 emission targets with environmental wellbeing, other and more devastating consequences of industrial processes are put in place as part of a 'carbon emissions reduction' programme. Paradoxically, while necessary in themselves, government projects focussed purely on cutting energy dependence from hydrocarbon fuels are counter-productive and bring greater harm to the marine balance, upon which most of the world's living organisms depend.

Waste dumps are found in all the five oceanic gyres, a full survey by Australian scientists finding that the Indian Ocean gyre contains a debris field of 2 million square miles (5km² million), with a full 360° rotation around the ocean taking six years before it spirals to the centre, where it will remain indefinitely. The Indian Ocean gyre is isolated from conjunction with the Atlantic and Pacific gyres and the collection of debris and refuse is less likely to bleed away into other seaways.

Cycles of change

The current period of change being forced upon the oceans by industrialisation is by no means the first in which they have been subjected to extreme conditions, although others are part of a natural cycle. Not least among the most dramatic were the episodes of glaciation, when the level of the world's oceans oscillated between extremes. In Chapter 4 we explored the icehouse world created by the Huronian glaciation more than 2 billion years ago. More recently, repeated cyclical glaciations have come and gone, each affecting the oceans as well as the land surfaces. But there have been dramatic shifts in the way seas, oceans and continental land masses have interacted over the Phanerozoic.

RIGHT The bathymetry of the Kerguelen Plateau in the southern ocean governs the flow of currents south of the Indian Ocean gyre.

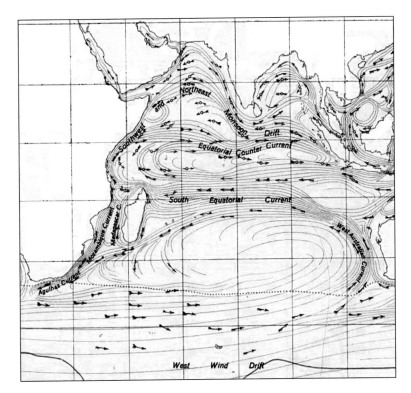

ABOVE The Indian Ocean gyre influences the flow of water from the southern ocean, but feeds a closed ocean compared to the North and South Atlantic Oceans and the Pacific Ocean. This focuses debris and rubbish dumped in the sea, leaving little opportunity for the dispersal of toxic waste, very little of which decays out. *(David Baker)*

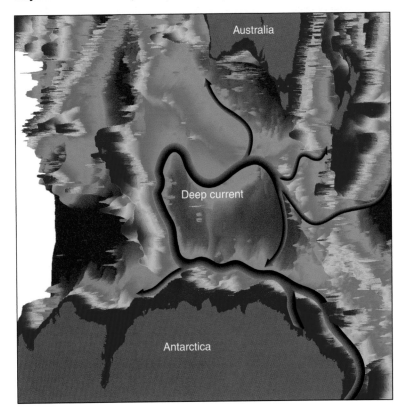

RIGHT The South
Pacific gyre follows a
path from the equator
westwards and down
to the East Australian
Current where it turns
to track the southern
ocean along the West
Wind Drift. Because
there is so little land in
the region of this gyre,
the ocean is depleted
in nutrients which
would normally flow
off rivers. *(David Baker)*

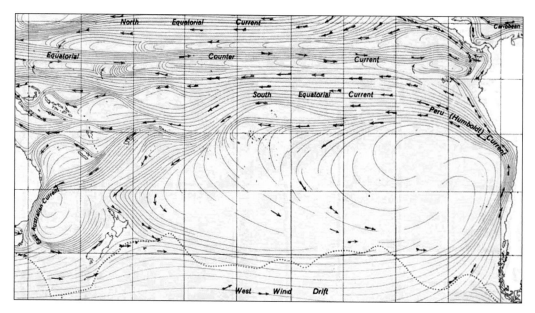

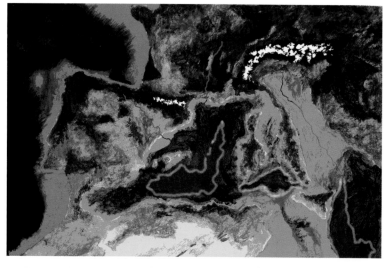

As the continents readjusted their positions
during the breakup of Pangea, the Tethys Sea
begat the Mediterranean, which closed up at
its eastern end due to the collision between the
Arabian microplate and Eurasia. By 5.96 million
years ago it was closed at its western end too
and the Mediterranean became an inland sea.
For 630 million years it went through a period
of partial opening north of the present Straits of
Gibraltar and entered a period of desiccation,
drying out completely halfway through this
period. Within 1,000 years the Mediterranean
basin was completely dry, a depressed
topography once part of the Tethys Sea, in
some places dropping down 16,400ft (5,000m)
below the then mean ocean level. It would have
been theoretically possible to walk from Africa

CENTRE What is now the Mediterranean,
formerly part of the Tethys Sea, closed up when
Africa pivoted round into Europe and landlocked
the sea at both ends. Large salt deposits
were laid down when the waters evaporated
and these have been explored and mapped
by oceanographers, who found remains of
land animals at the bottom of what is now the
Mediterranean. *(Roger Pibernat)*

LEFT About 5.3 million years ago the Atlantic
Ocean broke through, refilling the Mediterranean
and once again separating Africa from Europe at
the Straits of Gibraltar. *(Roger Pibernat)*

to Europe anywhere along the 2,200 miles (3,540km) separating the eastern and western ends of the old Tethys Sea.

And then, 5.33 million years ago, the Atlantic Ocean broke through the Straits of Gibraltar and cascaded down and across the vast plain with undulations and mountain ranges that would soon be completely covered with water, filling up the giant basin by more than 32ft (10m) a day. From the geological evidence it appears that the falls were relatively slow rather than a waterfall, but the slope would have poured water down a vertical elevation of more than 4,000ft (1,219m). According to assessments, the rate of discharge coming through the Straits was approximately 1,000 times the volume of water flowing from the Amazon into the Atlantic Ocean, which would put the flow rate at 7.381 billion cubic ft/sec (209 million cubic metres/sec).

Drainage from oceans into seas is not that common and the creation of the Mediterranean is an exception to the rule. But there are numerous major ocean basins fed by land drains that carry freshwater into the saline oceans, where sometimes fresh-water envelopes extend beyond the sight of land. One such is the world's largest, the Amazon basin in South America. Overall, 48.7% of the world's land drains into these basins and some scientists classify the Mediterranean as a drain for European rivers to the Atlantic. But the Amazon is impressive, Brazilian geographers claiming it to be the longest river of them all, at 4,345 miles (6,992km) only just beating the Nile river in Egypt and the Sudan, which is claimed to be 4,258 miles (6,853km).

For all the dispute over its absolute length, the Amazon basin is the largest in the world, with an area of 2.72 million square miles (7.05 million km²), discharging a total 55.2 million US gall/sec (209 million litres/sec) into the Atlantic. At its greatest time of flood, the Amazon is more than 30 miles (48km) across, with one side being out of sight to the other, internecine islands giving the apparent appearance of a riverbank. Each year the Amazon discharges 1,581.3 cubic miles (6,591km³) through the mouth of the main stem, which is 50 miles (80km) across. More than 1,100 tributaries enter the Amazon, each a river in its own right, and when it floods the Atlantic Ocean rushes in with

ABOVE Plunging 2,700ft (823m) in a single drop, Angel Falls in Venezuela is today the largest waterfall on Earth and resembles in some way the enormous fall of water which filled up the Mediterranean, flowing down a vertical distance of 4,000ft (1,219m). *(David Baker)*

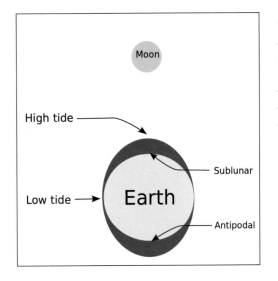

LEFT Both the Sun and the Moon are responsible for Earth's tides and this diagram shows the tidal influence of the Moon's gravity at the sublunar and antipodal sides of Earth. The Sun has a similar influence with a much smaller bulge at opposing sides of Earth in a direct line to the Sun. *(David Baker)*

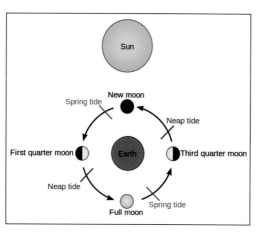

RIGHT Tides are at their highest when the Sun, Earth and Moon line up, which happens twice a month because those alignments occur at full and at new Moon, as displayed in this illustration. These times are said to bring spring tides, implying a bursting forth and not a season. The alignment itself is known as syzygy. When the Moon is at first or third quarter, its influence is much reduced, times that produce neap tides, meaning "without power". *(David Baker)*

waves 13ft (4m) high, flooding 8 miles (13km) inland across either bank.

Flooding is frequently the result of tidal flow and the primary cause of that is the Moon, which brings us back to the story of Earth's origin as a shared system between two worlds in co-rotation. Very early in Earth's history the Moon would have been much closer to Earth, only gradually drifting away and lengthening the day on Earth from four or five hours 4.4 billion years ago to its present 24 hours. The interaction between the Moon and Earth has been crucial to many of the geophysical events, as recorded in Chapter 3, because Earth's

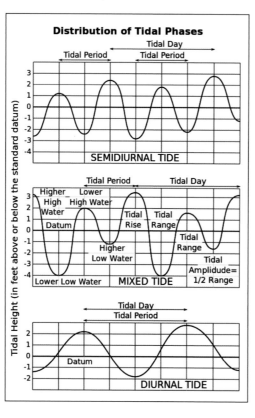

RIGHT There are various ways of measuring different types of tide according to conditions and location but the terminology remains the same. Diurnal tides have a frequency equal to Earth's day, or complete rotation. Semidiurnal tides occur twice a day. Mixed tides integrate several interactive forces to produce a variable range of tide heights. *(David Baker)*

engine has operated in conjunction with a gravitational exchange between the two bodies. But that exchange has not been equal.

The Moon is in an elliptical path around Earth, with the perihelion (minimum distance) being an average 225,320 miles (362,600km) and aphelion (maximum distance) an average 251,915 miles (405,400km), producing an eccentricity of 0.054. While Earth is tilted on its rotational axis 23.5° to its orbital path around the Sun – the ecliptic – the Moon's orbit is inclined 5.14° to the ecliptic, and, as we discussed in Chapter 2, it is locked in synchronous rotation with Earth. Complex motions dictate the way the Moon orbits Earth and how it precesses, like a spinning top, through a predictable cycle, but the dominant effect we are concerned with here is the gravitational effect on Earth's oceans.

Tides are in reality a side-effect of the gravitational force of one body upon another, defined by Isaac Newton's law of universal gravitation: $F = G\, m\, M_{Moon}/r^2_{Moon}$, where G is the gravitational constant, m is the mass of the object pulled by the Moon, M_{Moon} is the mass of the Moon and r_{Moon} is the distance separating the object (Earth) from the Moon. In order to understand the magnitude of attraction the Moon has for Earth, the distance between Earth and the Moon (r_{Earth}) together with the mass of Earth (M_{Earth}) can replace the lunar values in the preceding formula. Defining the force of Earth acting on the Moon (W) we can now write: $W = F = G\, m\, M_{Earth}/r^2_{Earth}$.

The mass of the Moon is 1/80th that of Earth. With the radius of Earth rounded out to about 4,000 miles (6,436km) and the Moon an average 240,000 miles (386,160km) distant, the distance of the Moon is 60 times greater than the radius of Earth. Because of the inverse square law, the attraction at the distance of the Moon on objects on or near the surface of Earth is 1/3,600. Because of the combination of these two effects, the Moon's pull on objects on Earth is a mere 1/300,000 of the equivalent weight on Earth. There is a measurable effect of this gravitational force on everything on Earth, water being the most obvious, with a graduated effect of the Moon's pull on one side of Earth and the other causing the rise and fall of tides.

However, it is not quite as simple as that because Earth rotates on its axis 27 times faster than the Moon moves around Earth, and that rotational torque drags the tidal bulges across the surface. Tides are also affected by frictional coupling with the solid surface of Earth, the inertia of the movement of water and by the gravitational attraction of the Sun, which is about half that of the Moon and which, added to the attraction from our companion, is responsible for spring and neap tides.

In addition, the Moon's gravitational attraction combined with that of the Sun causes tides within the body of Earth itself, with amplitudes of several yards, in a 12-hr cycle, but there are other daily and semi-annual tides that set up resonating oscillations between the movement of the water on the surface and the interior of the solid Earth. At full and new Moon times Earth, the Moon and the Sun are aligned, and this is where the maximum and minimum amplitudes appear when all three bodies line up, causing the greatest bulges on opposite sides of Earth. But the effects of this body-coupling is also observed in Earth's influence on the Moon.

The Moon experiences a tidal change of about 4in (10cm) every lunation (27 days),which combines a fixed influence due to the Moon's eccentricity and one from the Sun. Earth's influence is caused by the Moon's synchronous rotation and by libration, which is a combination of the eccentricity and the varying speed with which the Moon orbits Earth, which is faster at perihelion and slower (relative to Earth) at aphelion. This gives a strange visual effect where the Moon appears to wobble back and forth, allowing Earth-based observers to see 59% of the surface, while maintaining a fixed synchronicity with its parent body.

The tidal pull of Earth on the Moon causes seismic activity – moonquakes – measured by seismometers placed at four of the six lunar landing sites. They were operational from their date of emplacement (1969–71) up to 1977 when they were switched off, and a lot of data was obtained as a result of this extended period of observing moonquakes. Induced seismic profiling, where spent rocket stages were deliberately crashed on to the surface, allowed seismologists to study the internal structure of

the Moon, and it is from this activity that we know so much about its interior composition.

Moonquakes usually occur in two zones, more than 435 miles (700km) below the surface and within the first 19 miles (30km). These effects highlight the difference between Earth and Moon. Whereas the hydrous Earth damps out earthquakes relatively quickly, usually in a matter of minutes, the dry Moon can maintain shock waves from moonquakes for up to an hour after they are triggered. Other seismic events are caused by meteorite impacts, as they are on Earth when large bodies fall to the surface.

The oceans are a link to Earth's oldest geophysical events and are a reminder that this planet is but one in a two-body system driving dynamic forces above, at and below the surface. The oceans have been an integral part of the planet for all its evolutionary period and help control and manage the weather, the climate and the environment in which all living things thrive.

Summary

- Earth has had deep oceans and freshwater seas during almost all its evolution.
- Water behaves differently depending on depth, temperature and pressure.
- Variations in circulatory movement of water creates gyres.
- Coriolis forces determine movement of water in northern and southern hemispheres.
- Oceans are refuse dumps for artificial waste and industrial outflow.
- The Moon causes Earth's tides and Earth causes moonquakes.

ABOVE Understanding the balance between liquid and frozen water is key to understanding weather and climate on Earth. The European Space Agency's Cryosat satellite is the first dedicated to the observation and measurement of ice levels in Arctic and Antarctic regions. Launched in 2010 the satellite is a key tool in understanding ice levels as they change over time. *(ESA)*

Chapter Six

Evolution of atmosphere and environment

Earth has basked in a greenhouse environment dominated by high levels of carbon dioxide, dense atmosphere and high temperatures for 80% of its history. For the remainder of the time it has been an icehouse world, during periods defined by geologists and atmospheric scientists as ice ages.

OPPOSITE Blue light scattering at the top of Earth's thin atmosphere as viewed from the International Space Station, where the tenuous gaseous layer that surrounds the planet is host to life outside the oceans. *(NASA)*

We are in an icehouse world now, and some scientists believe humans are taking hold of this natural cycle and deflecting it from its logical progression toward another glaciation by pumping up Earth's atmospheric temperature, causing an enhanced greenhouse effect.

When Earth first formed and passed beyond the Archean to the Hadean eon, the atmosphere of the planet began to react chemically to the direct influence of solar radiation. The primitive Earth had no free oxygen, only that present in the mix of atoms coalescing around the protostar that would become our Sun. This allowed ultraviolet radiation to start photochemical processes that would involve water vapour (H_2O), carbon dioxide (CO_2), methane (CH_4), ammonia (NH_3) and hydrogen (H_2). The very early atmosphere would have been largely hydrogen and helium, as would most of the gaseous material in the evolving solar system. But not for long.

Over time the build-up of oxygen through photosynthesis in plants would create an ozone layer through the dissociation of the oxygn to create a protective canopy, beneath which complex living systems could evolve

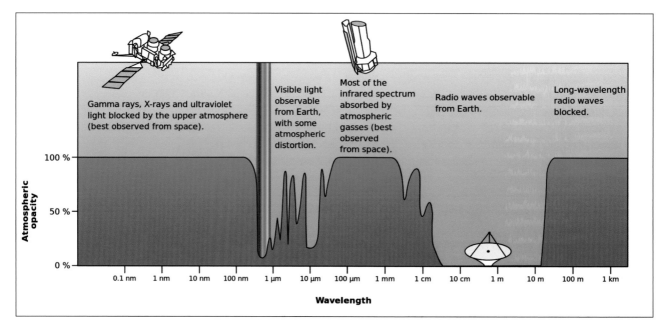

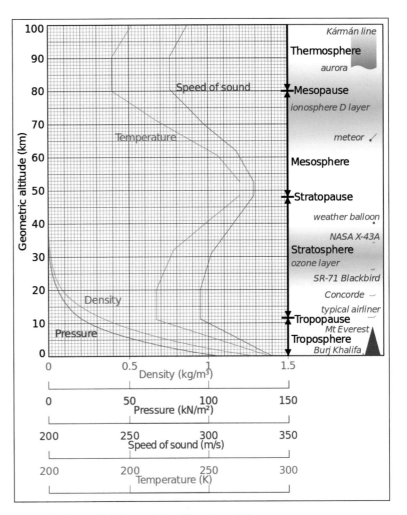

and proliferate. Meteorite impacts, violent thunderstorms and ultraviolet radiation reaching the surface of the early Earth, which was without an ozone layer, would have been critical to starting the photochemical reactions. Eventually, radiation would be a hindrance and not a stimulant to advanced life as it evolved, and the forming of an ozone layer was an important prerequisite for those events. But many changes were to take place before that could happen.

Primordial Earth

In rounded numbers, Earth's atmosphere today consists of 78.1% nitrogen (N_2), 20.95% oxygen (O_2), 0.93% argon (Ar) and just under 0.04% composed of carbon dioxide (CO_2), neon (Ne), helium (He), methane (CH_4), krypton (Kr) and hydrogen (H_2). It is an evolved environment produced by gases exhaled by a breathing planet and through photosynthetic activity and other biological processes from an increasing complex tree of life, diversified over 4 billion years. The atmosphere we breathe was toxic to life for more than half the history of the planet, and the gases it contains today are being modified by animal behaviour – a classification of life including humans – which is threatening to make the most fundamental changes to the atmosphere ever recorded over such a brief period of time.

Earth's first atmosphere was delivered by the solar nebula during the accretion process and would probably have consisted of largely hydrogen and helium, but there is no way of knowing for sure what the precise ratios

were. Jupiter, which has retained its primordial atmosphere, is 81% hydrogen and 18% helium with a few other gases. During the very early accretion phase some of the early atmosphere would have been lost due to high temperatures from incoming meteorites. Some of the energy absorbed by the planet would probably have been converted to gas atoms and these would be lost to space. Equally, the early atmosphere could have been added to by volatile elements that would begin the process of changing favourably towards the next phase.

Prior to the establishment of the magnetosphere, Earth's surface would have been swept by the solar wind, stripping away the attached atmosphere that was trying to build an envelope surrounding the planet. Helium and hydrogen would have been early candidates, drifting off into space and allowing a dramatic change in the composition of the gaseous canopy. Water exposed to ultraviolet radiation undergoes photodissociation to produce

molecular hydrogen, which then dissociates further into hydrogen atoms and escapes.

Clearly, as we will see from the record of Earth's history, its early atmosphere was quickly replaced with a denser and more enduring envelope. But the duration of this initial period of degassing can be traced through an isotope of iodine, ^{129}I, which is likely to have formed in a supernova explosion prior to the formation of our proto-Sun. Iodine-129 has a half-life of 15.7 million years, completely decaying out into ^{129}Xe, an isotope of xenon, in under 150 million years. If all the ^{129}I had decayed into ^{129}Xe the isotopic ratios from the interior of Earth after 150 million years should be the same as that of the atmosphere today. Which is not the case. An excessive quantity of ^{129}Xe

ABOVE The Aurora Australis as seen from the Amundsen-Scott research station in Antarctica, where we are reminded that the atmosphere, the magnetosphere and the solar wind are all connected as particles spiral into the magnetic cusp. *(Chris Danals)*

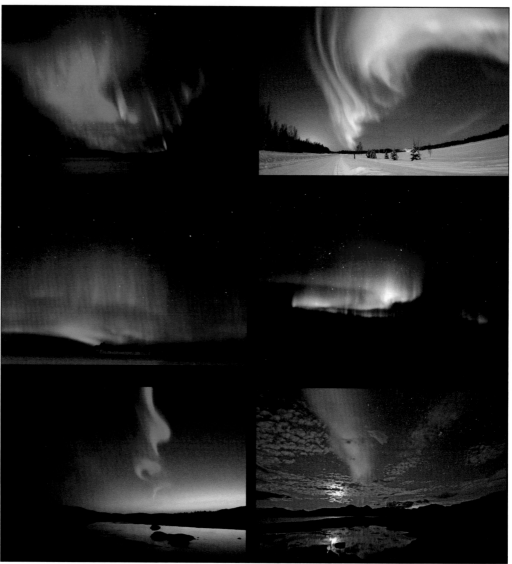

RIGHT A montage of views showing the Aurora Borealis over the Arctic region and the Aurora Australis in the southern hemisphere. *(14J Bella)*

in the mantle indicates that the degassing had been completed while [129]I was still around, so that puts a maximum limit on the date of the degassing and the assembly of a permanent and evolving atmosphere.

There is another way to measure the early atmosphere. The atmosphere today contains about 5ppm of helium of which the majority is [4]He and only around 7×10^{-6}ppm is [3]HE. The heavier isotope of helium is a product of the decay of radiogenic materials such as uranium and thorium, while the lighter isotope is mainly the result of the nucleosynthesis that occurred in star formation. Because helium only stays within the atmosphere, any traces of it today are of very recent origin. But the ratio of the lighter to the heavier helium isotopes at the bottom of the oceans is eight times greater than the ratio of those isotopes in the atmosphere. In places where the mid-ocean ridges are exposed above sea level, Iceland for instance, there is a 37:1 ratio. The greater abundances of the lighter [3]He isotope found at these extrusions are degassed products of the early atmosphere that are being brought from the mantle to the surface.

This ratio imbalance also holds true for ratios of neon ([20]Ne/[22]Ne), where a similar story unfolds when the isotopes are observed from the mid-ocean ridges and therefore emerging from the mantle. The ratio between the helium

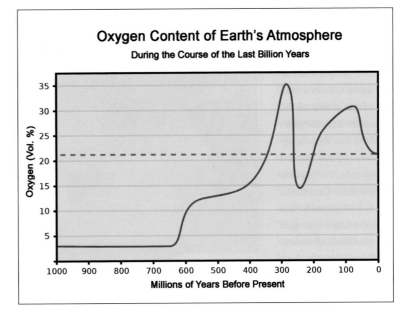

Oxygen Content of Earth's Atmosphere
During the Course of the Last Billion Years

ABOVE In much the same way that carbon dioxide levels have oscillated between very high and very low density when compared to present readings, oxygen levels too have fluctuated as a percentage of the atmospheric mix. Whereas CO_2 is a product of geological activity in Earth, O_2 levels are determined by living organisms. In this way Earth can truly be regarded as an integrated system, an engine continuously adapting itself. *(David Baker)*

BELOW A global map of water and moisture levels in the atmosphere from cloud observation and monitoring helps scientists understand the connection between seas, oceans and the atmosphere, much of this work being conducted on an international scale. *(NOAA)*

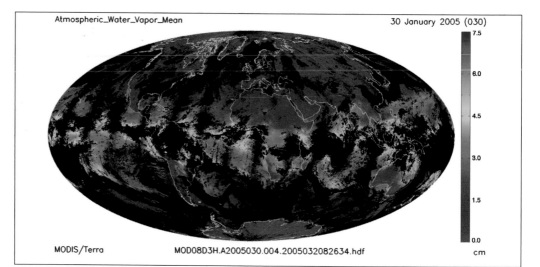

Atmospheric_Water_Vapor_Mean   30 January 2005 (030)

MODIS/Terra   MOD08D3H.A2005030.004.2005032082634.hdf   cm

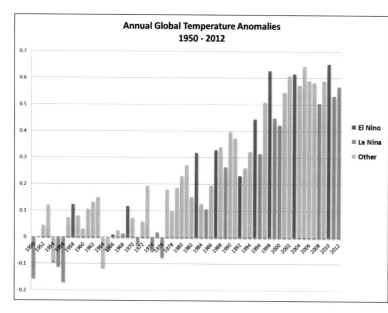

Annual Global Temperature Anomalies
1950 - 2012

■ El Nino
■ La Nina
■ Other

LEFT Atmospheric events are recorded in temperature readings since 1950 where anomalies at variance with the mean average are shown for El Nino, El Nina (a variation on thermal balance) and other phenomenon. The clear indication is that temperature deviations are getting more extreme. *(David Baker)*

to a period between 10 million and 70 million years after the beginning of the accreting Earth. Volatiles introduced from meteorite impacts and the occasional comet are a key to defining the nature of the early Earth, and to fully understand the magnitude of changes which are believed to have occurred in the makeup and changing pressure levels it is necessary to use a comfortable terminology.

Measuring pressure can be complicated. There are so many units of measurement in use. Today the sea-level pressure of the atmosphere (atm) is approximately 760mm of mecury, or $14.7lb/in^2$ (101.325kPa), fractionally above 1 bar (which is 0.987atm). A bar is a unit of measurement equal to 100,000 pascals,

LEFT Arctic sea ice has had a controlling influence on the use of the polar region for maritime operations, closing out quick routes between Europe and North America to the Far East. The consequences of diminishing sea ice is a constant concern to environmentalists, however, since a decrease in sea ice attracts commercial activity, including drilling for oil. *(David Baker)*

or 100kPa. When measuring large shifts in atmospheric pressure it is helpful to use the abbreviation atm, much as a distance scale in the solar system used the AU (astronomical unit), the distance from the Sun to Earth, as fractions or multiples for the relevant location of planets, etc.

Greenhouse world

If all the water on Earth and in the oceans were present in the form of vapour, atmospheric pressure would be 270 bar, and if all the carbon dioxide in the planet were to be placed back out in the atmosphere the pressure would be a staggering 480 bar. Moreover, after the beginning of the accretion process and differentiation of metallic core and silicate mantle the surface temperature would be around 3,632°F (2,000°C) due to the broiling planet bombarded with meteorites. On calculation, pressure would fall to a third of that after 4 billion years. But that is not unreasonable, given that today Venus, our planetary twin, has a surface pressure of 90 bar and a temperature of 869°F (465°C). It is, therefore, reasonable to say that much of the water Earth was ever

ABOVE An iceberg in the South Atlantic Ocean, one of many logged and monitored for safety of shipping and for details allowing scientific investigation of ice mass distribution analysis to be assembled. *(NASA)*

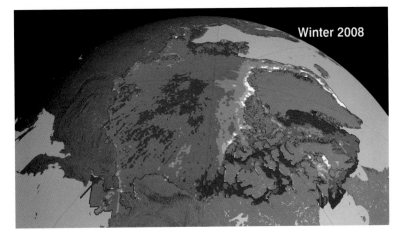

Winter 2008

RIGHT Arctic sea ice levels in the winter of 2008. Spitzbergen is between Norway, the land mass at top right connected to northern Russia as it wraps around this spherical projection, with open water (light blue) to the south and sea ice around its northern and eastern areas. *(NOAA)*

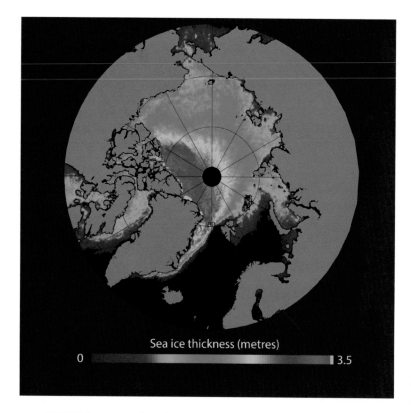

ABOVE In averaged sea ice measurements between 2010 and 2015, the thickest is found north of Greenland and north-eastern Canada. Iceland is the island at the bottom between Greenland and Norway, with Svalbard islands containing Spitzbergen between the northern tip of Norway and the North Pole. *(NASA)*

BELOW Glaciers and ice-world conditions have characterised large periods in Earth's history. This chart shows the 4.6-billion year story of Earth with long periods in which partial or total glaciation of the planet was normal. Note too that the Great Oxygenation Event was coincident with major climatic change across the globe. *(USGS)*

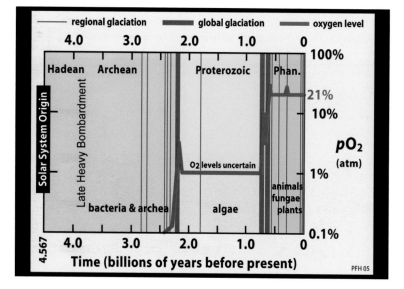

going to get had been acquired by at least 4.5 billion years ago, when pressures were probably 310–480 bar.

The next phase would be rapid cooling of the surface to allow a silicate crust to form, previous temperatures (above) being far too high for the outer materials to solidify. Gradually, over the next 30 million years, the pressure of the water in gas form remained at 270 bar while the pressure of the carbon dioxide went down to 40 bar. With the surface temperature falling, the very strong greenhouse effect delayed solidification of the crust, but by 4.4 billion years ago the water vapour had condensed down into seas and oceans, the CO_2 pressure had fallen to 40 bar and the surface temperature had reduced to 572°F (300°C). This sequence suddenly opened the possibility of a steaming greenhouse environment where temperatures at the surface remained at around 392–482°F (200–250°C), conditions that would set up the first permanent atmosphere and establish a water-world planet.

With the establishment of a tectonic cycle and the early formation of a crust, the atmosphere was largely carbon dioxide with molecular hydrogen, molecular nitrogen and methane. With a surface pressure at least 40 times that of the present atmosphere, high temperatures in the oceans forced leaching on the sea floor producing silica and bicarbonate in this reaction: $MgSiO_3 + 2(CO_2) + H_2O \rightarrow Mg^{2+} + 2(HCO_3^-) + SiO_2$. In this way magnesium silicate produced bicarbonate and silica, the bicarbonate HCO_3^- precipitating out as calcium carbonate $CaCO_3$ in the following sequence: $2(HCO_3^-) + Ca^{2+} \rightarrow CaCO_3 + CO_2 + H_2O$.

In this manner carbon dioxide from the atmosphere, together with that dissolved in the oceans, became trapped in the carbonate minerals. It was because of the presence of liquid water that this cycle of CO_2 burial began to take hold and with it the reduction of atmospheric pressure as it was being captured in rocks taken down into the mantle and held there. The effect was to chill down Earth and reduce the warming effect of a greenhouse environment. This runaway decline in CO_2 content may have reached levels where atmospheric pressure fell to only 20% of its present value, only to rise again when the process came to a stop, as a frozen Earth would be replenished with more carbon dioxide from volcanic activity. As the pressure increased so would the temperatures, starting a renewed cycle bringing successive waves of global glaciation.

By whatever mechanism, evidence from the oxygen and silicone isotope measurements in sedimentary rocks from the boundary of the Archean and the Hadean eons indicates a fairly constant temperature of 158°F (70°C). This would imply a carbon dioxide atmosphere of about 3 bar, a surface pressure three times that of Earth today. From this point on the emergence of life played a major role in modifying the atmosphere, and this has remained so to the present day. Much of that has to do with the amount of hydrogen in the environment, because it has a reducing effect on the atmosphere, and that in turn encourages the synthesis of organic molecules.

As molecular hydrogen is released from basalts in the ocean crust as a product of their alteration by high-temperature water in the hydrothermal vents, it is liberated to the atmosphere where some gets lost out of the top of the atmosphere while some is consumed in prebiotic chemical reactions. These reactions would have produced hydrogen sulphide (H_2S), with methane also playing a significant role in changing the atmosphere. Because the escape of free hydrogen to space is easier the higher the temperature, the early outer atmosphere would have had almost no oxygen and would therefore have been cooler, slowing the loss and achieving a balance between that being produced out of the vents and the amount being lost to space.

Prebiotic chemistry would have absorbed a lot of this hydrogen and, along with other reducing agents, set the scene for the production of organic molecules. The presence of these other gases is quite important for setting up the conditions known to exist at the early stages in the formation of life. For

ABOVE Earth's northern hemisphere as it would have appeared during the last major glaciation. Most of the Mesolithic populations of northern Europe would have remained in their familiar hunting grounds, which was shielded from the worst effect of the ice. *(NASA)*

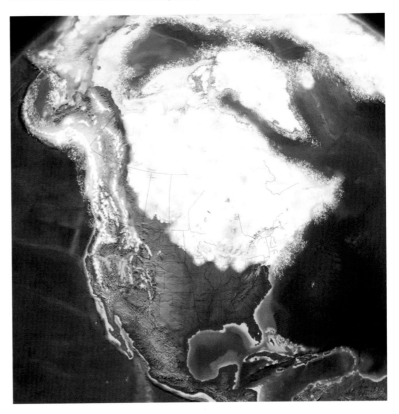

RIGHT Across North America, up to 12,900 years ago the Younger Dryas covered most regions down to New York and Baltimore with the area immediately east of the Rockies free from ice, the route from Asia to the North American continent accessible by coastal zones. *(NOAA)*

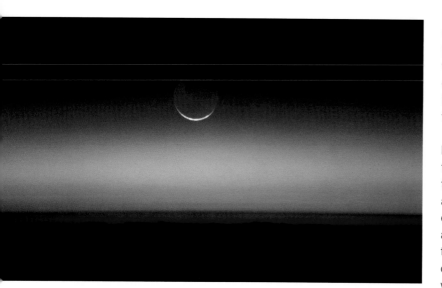

instance, there is uncertainty about the amount of nitrogen in the atmosphere, a reducing gas that may have been at its present level of 0.8 bar though there is also evidence to show that it might have been two or three times as high, and that this alone would have stirred up the conversion of organic nitrogen into N_2.

The presence of methane is an important regulator of the greenhouse temperature, being very much more efficient at that than carbon dioxide. This alone could have helped regulate the temperature of Earth from swingeing and highly damaging cycles between extreme cold and intense warming, and that would have been a very important factor in stabilising Earth for the development of life. Moreover, in the absence of dioxygen (O_2) the methane would have had a molecular lifetime up to 1,000 times greater than it has today in a relatively oxygen-free

environment. Possibly boosted by methanogenic prokaryotic organisms the production of methane would have had a cooling effect on Earth too, by creating an upper stratospheric haze that would have increased the reflectivity of the atmosphere to solar radiation.

Nevertheless, a balance emerged which kept the surface and lower atmosphere warm through the highly efficient means of raising the greenhouse environment while reflecting away sunlight. It appears likely that this anoxic environment, in which methane could retain a stronger controlling influence on both the temperature of Earth and the production of organic compounds, was an essential base from which dramatic changes to the atmosphere were about to take place. But it was a comparatively unstable situation and one in which a delicate balance existed between greenhouse-world and icehouse world. The great glaciation of 2.9 billion years ago is an example of how unbalanced that divide could be.

About 2.3 billion years ago Earth was subjected to a sudden accumulation of oxygen in the atmosphere at an unprecedented level. It used to be thought that Earth had a little oxygen in its atmosphere, together with carbon dioxide, from which the oxygen production was linked directly to the proliferation of life. Yet as we will see later, cyanobacteria had been around for about 200 million years before this event, described in Chapter 7 by its somewhat unimaginative name of the Great Oxygenation Event (GOE). It is also referred to as the first great extinction event. While it is

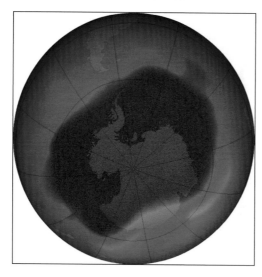

probable that cyanobacteria produced the first concentrations of oxygen through oxygenic photosynthesis, free oxygen would have rapidly been absorbed into organic matter or dissolved iron, maintaining an anaerobic environment that suited life at that time.

Because free oxygen is a poisonous gas to anaerobic life, so long as it remained in the sinks – accumulating all the time – various forms of such life would have been safe. But when the oxygen sinks became full the free oxygen reacted with methane in the atmosphere, which significantly reduced its concentration, lowering the greenhouse effect that triggered the Huronian glaciation described in Chapter 4. Because solar output was significantly lower than it is today, the average temperature across Earth was sufficiently high to prevent the oceans freezing over because of the greenhouse gas methane, maintaining a comfortable temperature and keeping the fluid (water) in its liquid state.

Some scientists believe that a reduction in carbon dioxide levels at this time could also have contributed to a reduction in the greenhouse effect, although CO_2 is weak as a greenhouse gas compared to methane and this contribution may have been marginal at best. Nevertheless, the presence of free oxygen, while a poison to anaerobic life, was to significantly influence the settlement of the atmosphere and the emergence of aerobic forms. There are some questions about the emergence of oxygen that fuel debate, some scientists believing that dioxygen was locked up in Earth and liberated by geophysical processes, but this does not bear scrutiny because the only believable mechanism for producing oxygen in these quantities is photosynthesis.

Turbulent times

For more than 2 billion years photosynthesis and oxygen production have gone hand in hand. Oxygen levels peaked at 35% of atmospheric volume between 95 million and 68 million years ago, having increased steadily since 2.3 billion years ago. There had been fluctuations but these were temporary in comparison with what is happening today. Oxygen levels had been steadily falling since

ABOVE Triggered by a westward moving wave off the west coast of Africa, tropical storm Marie developed off the west coast of North America in August 2014 and became one of the most destructive hurricanes ever recorded in the Pacific Ocean. Winds of 160mph (260kph) were noted. *(NASA)*

BELOW The first hurricane recorded in the 2004 season, Alex developed from a high-level depression and a weak surface trough to move up the eastern seaboard, threatening US cities as it moved north and out into the Atlantic Ocean. Hurricanes are given names in alphabetical order of their emergence and the same name can be used for several different years. *(NOAA)*

ABOVE A spectacular view from space reveals the intensity of Hurricane Amanda in May 2014, its eye clearly discernible. As it intensified, pressure dropped by 57 millibars in 24 hours, but fortunately the storm did not encroach on land. *(NOAA)*

the great extinction in the Cretaceous-Palogene boundary 65 million years ago, but as related at the beginning of this chapter they are now down to a mean average of 21%.

There are many reasons why oxygen levels are falling. Forests and phytoplankton, so important for life-supporting oxygen production, have been under attack for decades as managed rural landscapes are transformed into stripped prairies for food production on a colossal scale, consumption demands driven by an almost exponential population explosion. Every day, 50,000 hectares of oxygen-producing forests are being cut down, while massive deforestation is taking place legally. And as the expansion of industrialised

agriculture and grazing land has exploited naturally evolved agrarian sites, artificial fertilisers are forcing reactive nitrogen that is binding oxygen in plant tissue, in soil organics and into the ocean along with nitrates.

In cities where pollution is a product of urbanisation and dramatic overpopulation, where dense populations of 20 million and more are not unusual, it is not only pollution and toxic chemicals that threaten the balance of life. It is also in the destructive effect of coal-burning and unconstrained hydrocarbon consumption that pushes oxygen levels down in some dense urban areas. More than any other animal, humans require 25% of their energy intake just to run the brain, compared to 9% for a chimpanzee. Any significant lowering of oxygen affects the human brain, and neuroscientists are increasingly concerned at the debilitating effect of oxygen starvation. Significant increases in cancers and other so-called 'modern' diseases are a direct product of densely urbanised regions where levels of pollution have been increasing decade upon decade.

It has been estimated that the expansion of the global population, and the attendant impact which that alone is having on the atmosphere – not counting any influence on climate or Earth's temperature – will cause clinically life-threatening debilitation when the population reaches 12 billion people. It has expanded from less than 2.5 billion to more than 7 billion in the 70 years since 1945. Little of this is even being considered when urban growth is frequently hailed as a wondrous hallmark of human progress, and the rights of humans to proliferate are an unquestioned hallmark of free will. But there are other threats to our idyllic picture of a benignly tranquil Earth.

In Chapter 3 we saw the effect impacting meteorites can have on the geological record,

LEFT Radar maps track hurricanes and tropical storms, informing weather prediction specialists so that alerts can be sent to local emergency services and news networks. Hurricanes incur substantial damage and loss of life. Weather forecasts are vital in many tropical and subtropical regions for implementing lifesaving and damage-limitation activities. *(NOAA)*

leaving an indelible imprint of catastrophic events on a colossal scale. But while it was said that such effects are a blip on the overall evolution of the planet, a much more enduring and permanent change can be the result of such events on the way they affect the atmosphere, climate and weather. Volcanic activity itself can change dramatically the way an ecosystem is evolving – much more so the great plume events that created the Siberian and Deccan traps, to name but two in a series of such extrusions occurring during the Cenozoic – the last 66 million years. Giant impacts can effect change on an even greater scale, and much research has been carried out on the very real influence these have on the atmosphere, sometimes with alarming results.

Traps are not only found on continental plates: several have occurred on ocean crust, on which they are known as plateaus, examples being (with ages in brackets) in Java (120 million years), Kerguelen in the Southern Indian Ocean (118 million years), the South Atlantic (80 million years), the component of the Deccan trap offshore in the Indian Ocean (66 million years), and the North Atlantic (56 million years). The great quantity of material breaching extant ocean crust changes the geochemistry of the sea floor and this can be seen in marine sediments that contain the fossilised record of the world's oceans, where the vast majority of living organisms can be found.

Clathrates are chemical substance that trap molecules in lattice structures, most notable among them being methane, where the preference for hydrogen bonding is met with water (H_2O). But other molecules of a trapped gas include carbon dioxide (CO_2), hydrogen sulphide (H_2S) and methane (CH_4) among others. Clathrates store chemical compounds that can otherwise adversely affect an environment. For instance, the release of 35.3ft³ (1m³) of solid methane clathrate would release 5,936ft³ (168m³) of methane gas. Locked in deep water under extreme pressure in the world's oceans, methane reacts with the hydrogen in the water to form solid methane clathrates.

Methane is 23 times worse than carbon dioxide as a greenhouse gas. Were it to be released the overwhelming effect on the atmosphere would be to tip it into a period of

warming far beyond anything predicted by the wildest doomsayer on climate change. It is kept intact on the ocean floor and in dewars suspended on the continental shelves, and the partial release of significant quantities of methane clathrates in former times when great climatic upheaval was experienced can be explained by this mechanism. If the world warms up – for whatever reason – the methane clathrates begin to dissolve and release methane as a gas. This has, in the past, been highly effective in the seesaw rise and fall of climatic conditions.

In the Permian extinction the temperature of Earth increased by 9°F (5°C). This resulted in the release of methane, with some carbon dioxide, which raised the temperature by the same number of degrees as a result of the greenhouse effect. And the destabilisation of the methane clathrates caused widespread damage to climate and the biological processes, deflecting the course of evolution. The effect was exacerbated because only a modest increase in temperature can cause magnified repercussions.

Evidence for this comes from the analysis of carbon isotopic variations of ^{13}C found in sediments, and an observed increase in the abundance of ^{12}C at the end of the Permian shows this to be related to the extinction event. Release of ^{12}C-rich clathrates would trigger the release of large quantities of methane into the atmosphere, decreasing the $^{13}C/^{12}C$ ratio and consequently the measurement of the isotope ^{13}C – exactly what was found.

The traps and plateaus which become destabilised at times of great upheaval, caused

BELOW Hurricane seasons are mapped and tracked with great detail to inform computer models on future consequences of these advanced meteorological phenomenon. This tracking map shows the 2014 hurricane season across the eastern Pacific Ocean. (NOAA)

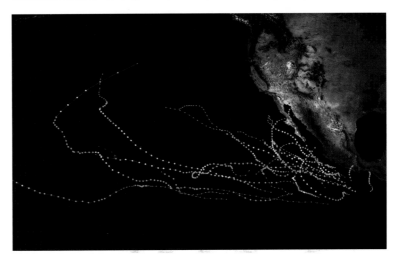

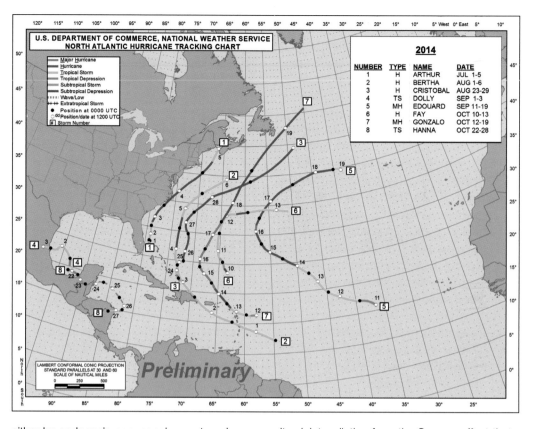

U.S. DEPARTMENT OF COMMERCE, NATIONAL WEATHER SERVICE
NORTH ATLANTIC HURRICANE TRACKING CHART

2014

NUMBER	TYPE	NAME	DATE
1	H	ARTHUR	JUL 1-5
2	H	BERTHA	AUG 1-6
3	H	CRISTOBAL	AUG 23-29
4	TS	DOLLY	SEP 1-3
5	MH	EDOUARD	SEP 11-19
6	H	FAY	OCT 10-13
7	MH	GONZALO	OCT 12-19
8	TS	HANNA	OCT 22-28

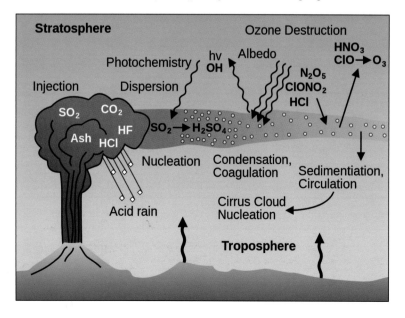

either by endogenic or exogenic events, release large quantities of hydrogen sulphide (H_2S) and sulphur dioxide (SO_2), which mix to produce sulphur trioxide (SO_3) and combine with water to produce sulphuric acid (H_2SO_4), which falls down on the surface of the continents and on the oceans as acid rain. As they reach the atmosphere these gases react with the ozone layer, depleting it and admitting high levels of ultraviolet radiation from the Sun, an effect that takes several thousands of years to reverse.

The effects of these geological and impact events have long-term consequences, and Earth never quite goes back to the place where it was before the occurrence. The Siberian traps were responsible for an increase in Earth's temperature to an estimated 40°C (about 105°F) 250 million years ago, clearing Earth of the vast majority of its living organisms and changing forever the evolution of the planet. In fact, life forms have frequently been contained within periods of great convulsion separated from each other by several hundreds of millions of years, and it is a misconception to believe that any mitigation in climatic change will restore it to its former condition, somehow reversing the clock and restoring prior conditions.

When discussing volcanoes in Chapters 3 and 4 we saw how effective Earth is at modifying the atmosphere through volcanic eruptions, billions of tonnes of magma, ash and dust particles discharged into the air, frequently to very high altitude. Rapidly solidifying molten particles explode without the constraint of the pressure in the dome from which they came, and together with pockets of gas form glass, created when the atoms in the magma have insufficient time to organise into molecules by forming a lattice structure. The random arrangement of the atoms in the glass spherules ejected to extreme altitude, extending high into the stratosphere, circulate on winds that envelop the planet and can remain aloft for more than a year.

Heavier particles will rain down quickly, intermediate-size fragments taking days or weeks to descend, causing reduced sunlight to reach the ground and cooling over wide areas. We have already seen how massive eruptions in the past have cooled the planet for several years before the reversal due to warming caused by higher levels of carbon dioxide in the atmosphere. When ash is the main component liberated from the volcano, the effect is less than when large quantities of sulphur are released. Sulphur compounds combine with water to make sulphuric acid and these droplets absorb and scatter light, but because the stratosphere is very dry it can take a very long time for that to happen.

In recorded times, it has been found that heavy emissions of sulphur droplets can cause cooling for up to two years, and it is these particles that pose the greatest hazard to life. When it exploded in 1991, Mount Pinatubo in the Philippines liberated 15 million tonnes of sulphur dioxide into the atmosphere, and satellites tracked the cloud in the stratosphere as it reduced global temperatures by 1°F (0.6°C). Even a moderately sized eruption can cause devastation to climate balance, as evidenced by the Little Ice Age that lasted for several hundred years as a result of four eruptions between 1275 and 1300. Atmospheric aerosols prevented the usual summer warming periods from taking hold across the Northern Hemisphere, which weakened the Atlantic Ocean currents, shut down the transfer of heat and expanded Arctic ice fields.

As evidence grows supporting the finely tuned effects of volcanic emissions and their interaction with the atmosphere it is possible to see where short fluctuations in temperatures and atmospheric conditions over just a few decades can be attributed to eruptions perhaps a long time before the effect is registered. Feedback mechanisms can amplify the

effects of these changes to the particulate and chemical balance of compounds in the atmosphere. There is growing uncertainty about assumptions made in attributing effects of what we measure in the atmosphere to what we may wish to interpret as obvious causes.

There is a chilling flaw in the human psyche: if we find a solution we think we can control, we adopt that rather than face inevitable challenges to us as decision-making people – that what we face in the future is probably unstoppable, given our inability to change the root causes: overpopulation, conspicuous consumption, and a hefty chunk of denial.

Summary

- Earth's atmosphere has been transformed by intensive bombardment.
- Carbon dioxide has been at an historic high for most of Earth's recent geological past.
- Oceans play a vital role in regulating carbon dioxide levels.
- The rapid build-up of oxygen during the second half of Earth's history changed the climate.
- Unstable geological activity can induce catastrophic changes to the atmosphere.

ABOVE Japan's meteorological satellite Himawari observes the Earth's weather from a fixed, geostationary orbit 22,300 miles (35,900km) above the planet. Data is also gathered by satellites from the European Space Agency, India, Russia, and the United States, the host countries providing information that is shared universally and made available without charge to all nations. *(JAXA)*

Chapter Seven

The impact of life on Earth

The biggest question faced with an uncertain, perhaps unknowable, answer concerns the origin of life. There is no direct evidence for when life arose on Earth and this planet is the only one known to have supported the development of living organisms.

OPPOSITE The Jurassic world of 150 million years ago saw the largest land animals of all time roam majestically through vast expanses of vegetation. It marked the spectacular peak of sauropod development. *(Gerhard Boeggemann)*

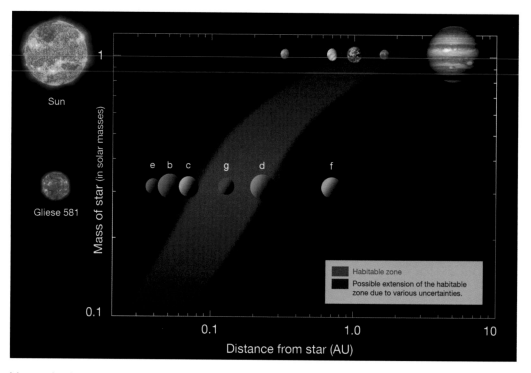

RIGHT The desire to understand the origin of life sends unmanned expeditions to the solar system in search of clues to how it might have begun. It also searches for solar systems which may contain Earth-like planets suitable for life. One such is Gliese 581 which is a relatively small star but with planets known to be in its eco-zone, a distance where life could thrive and water could exist in liquid form. As shown here, one planet (g) lies within the star's habitability zone. At top is the habitability zone for our Sun, with Earth and Mars within the suitable range. *(NASA)*

Many scientists remain convinced that life is prolific in the universe and that it is only a matter of time before primitive, non-intelligent, life is discovered in this solar system. Or at the very least, some evidence, perhaps on Mars or some of the moons of the outer giants, that life did exist at some point in the past when conditions were favourable for its evolution.

The question itself is without proper definition because there is no universal agreement as to what constitutes life. There is agreement, however, that organic molecules are found in space and on planets and that they could have arrived on Earth as a natural process of accretion followed by bombardment from meteorites, or that they could have organically synthesised through ultraviolet light or electrical discharge from violent thunderstorms. Moreover, the presence of organic molecules may not be linked to the emergence of life. It used to be thought that anywhere organic molecules were found, life was sure to follow. But that belief is now fighting a rearguard action against powerful scientific argument. It is linked more through association than a pragmatic evidenced-based conclusion.

The presence of organic molecules belongs in the realm of prebiotic chemistry, also known as abiogenesis when relating it to the origin of living things. Organic matter is an implicit ingredient in the interstellar medium – life is certainly not. However, these are the essential building blocks and the conditions for life to emerge may have appeared twice on Earth: before and after the Late Heavy Bombardment period, lasting more than 100 million years, about 600 million years after the formation of

LEFT An artist's impression of what a trinary star system with an Earth-like world might look like, envisaged as it might be at the beginning of its evolutionary form and before life emerged. *(ESO)*

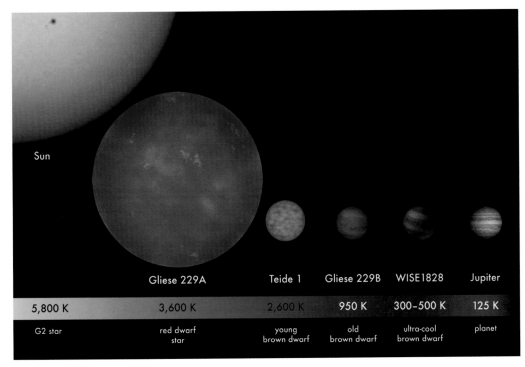

Sun	Gliese 229A	Teide 1	Gliese 229B	WISE1828	Jupiter
5,800 K	3,600 K	2,600 K	950 K	300–500 K	125 K
G2 star	red dwarf star	young brown dwarf	old brown dwarf	ultra-cool brown dwarf	planet

the solar system and centred about 3.9 billion years ago. Solid evidence for the existence of life can be found in biogenetic graphite from western Greenland dating back 3.7 billion years and in microbial mat fossils from 3.48 billion years ago. Whether life existed before the LHB is problematical but there is some circumstantial evidence to support the view that it did.

Windows for the emergence of life appear twice, therefore. Physical, chemical and geological conditions for the appearance of biopolymers (biological assemblages of large molecules) and nucleic acids (very large polymers) for between 4.4 and 4.2 billion years ago are known to have existed. Whether they actually did or not is problematical. The formation of any of the interactive organisation essential for life would have to have been in the presence of water and within a temperature band of 32–212°F (0–100°C). These conditions could have existed before the LHB and certainly after, but it is not known whether there was a bridge for some strains of life to survive the intense reheating of the outer surface of Earth or whether life became extinct and began again.

The habitable zone is often quoted by astrobiologists, studying the possibility of life in other stellar systems as well as our own solar system, as being a favourable location close to a life-supporting star in which living things can evolve. The habitable zone of long-lived, cool stars lies close in, as they provide much less light and radiation. While being long-lived and therefore affording plenty of time for life to develop, their habitable zone lies within the zone of synchronous capture – one face of an orbiting planet permanently locked toward its parent star, boiling on one side, frozen out on the other. Alternatively, very hot stars have short stable lives allowing insufficient time for life to evolve.

Earth lies in the sweet spot of life-supporting properties around our Sun; although conditions

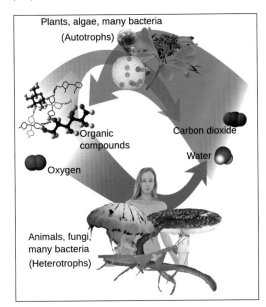

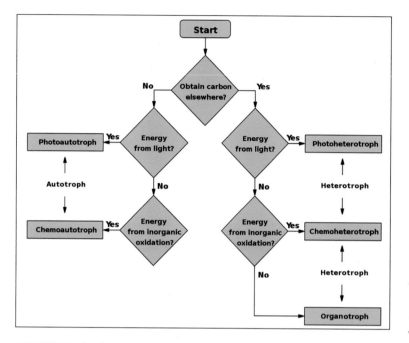

LEFT The definition tree with yes/no gates for determining what is an autotroph and what is a heterotroph. *(David Baker)*

were not commensurate within the habitable zone, Earth's surface was made more acceptable for life by the greenhouse effect that has dominated our planet's history. Without that, as we have seen, the world would have been permanently frozen until the increasing energy from the Sun made conditions more acceptable. It is not, therefore, the theoretical habitable zone alone which can dictate where life is possible around a star but rather the conditions on the surface of the planet.

It is a great mistake to take the conditions on Earth today as an analogue for life elsewhere, for we are merely defining carbon-based life – the one we know best – rather than silicon-based life which is theoretically possible but not very plausible. While possessing similar chemical properties to carbon, silicon is in the same group in the periodic table and can create molecules capable of carrying information necessary for biological life. It does have limitations, however, not least its inability to bond with a wide range of atoms necessary for metabolising.

Returning to the burning question 'What is life?', a useful starting point is the cell, the essential constituent of all living things whose function relies on carrying information that is simultaneously expressed, so that it can fabricate its own components, and on transmitting that package to form other cells. The storage, transmission and expression of genetic information requires a metabolising influence, but where that began is highly debatable. Nevertheless, the cell is the basis for the tree of life, comprising prokaryotes, eukaryotes and archaea, which are three categories of species.

The prokaryote is a single-celled organism devoid of separate cell compartments and without a nucleus, which is the most important difference between them and eukaryotes. They have a nucleoid that contains a single DNA (deoxyribonucleic acid) molecule carrying genetic instructions for reproduction. Prokaryotes also carry a tail-like flagellum that aids in cellular locomotion. Bacteria and archaea are prokaryotes but they are inherently

BELOW The food chain separates consumers into primary (herbivores feeding on plants), secondary (carnivores feeding on animals and some omnivores) and tertiary customers (feeding on both types from the top of the predatory tree). The feeding chain energises life, recycles product and permits evolutionary growth and expansion of species as well as species variation. *(David Baker)*

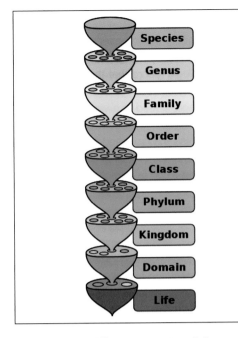

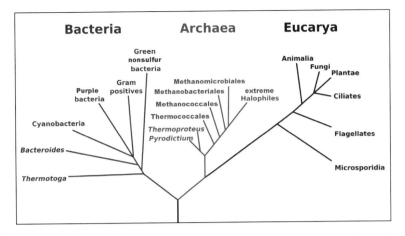

limited in their ability to support evolutionary development into plants, fungi and animals. The smallest of all organisms, prokaryotes are no more than 0.2–2.0µm in diameter.

Uniquely, eukaryotes have membrane-bound organelles, the life support structure for the cell, and a nucleus that contains the genetic material in the form of long linear DNA with proteins to form chromosomes containing the nuclear genome. While eukaryotes are the most apparently abundant structure for living things, in reality they are only a tiny fraction of living cells. In the human body, for example, there are ten times as many microbes as there are living cells.

There has long been debate about which came first – the apparently primitive and less organised prokaryote, or the more complex, seemingly more developed and certainly more evolutionarily friendly eukaryote from which we, as humans, are assembled. Some biologists believe that prokaryotes evolved into eukaryotes and that they are not competing, or parallel, evolutionary strains. One enduring truism in biology is that alternate routes and paths will eventually be sought and that this is the evolutionary method for diversification, even from prokaryote to eukaryote, supplementation with replacement – the parallel existence of both.

One explanation along this line of evolutionary transformation is that the prokaryote grew in size and developed folds on the interior face of the cell membrane to

LEFT The classification of living organisms is arranged in increasing orders of specific characteristics comprising eight major taxonomic ranks. Without an as yet unimpeachable definition, scientists move up the chain until separate species within a genus define highly channelled typology. *(Peter Halasz)*

ABOVE A phylogenetic tree of life based on the ribosomal RNA genes according to the three domains of life: bacteria, archaea and eukaryote. *(David Baker)*

BELOW Deoxyribonucleic acid (DNA) is the molecule which contains all the genetic instructions for development, function and reproduction of all living things and also of some viruses. Most DNA molecules are comprised of two biopolymer strands coiled together to form a double helix. The structure of the double helix shown here displays colour coded atoms according to element and the detail on base pairs is shown at right. *(Zephyris)*

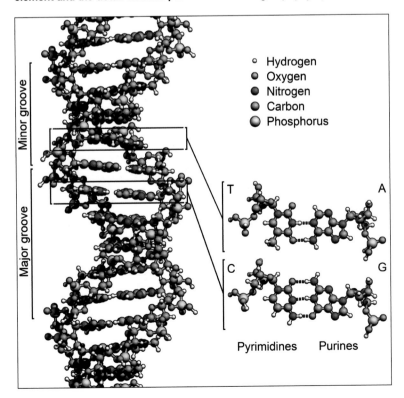

ABOVE The Ordovician world 460 million years ago saw a flourishing of life, with molluscs and arthropods (invertebrate animals) dominating the seas, along with fish and the very first vertebrates, prior to a mass extinction at the end of the period. *(David Baker)*

BELOW Eurypterus, common in the Silurian seas 430 million years ago, was the size of a large child and had a broad paddle-like arms for swimming and an extended tail as a rudder. They are thought to have predated or scavenged for food. *(Dimitris Siskopoulos)*

Earth's resources and began to systematically change the environment. Paradoxically, perhaps it was the appearance of the eukaryotes that gave a spur to prokaryotic diversification into bacteria, parasitically invading the more sophisticated structures and eventually the bodies of animals, with mixed consequences – pathogenic bacteria such as tuberculosis being but one example.

Once evolved, the explosive radiation of the eukaryotes produced a rich and diverse range of groups, within which a wide variety of types can be found. So abundant were they that scientists have difficulty arranging a phylogenetic relationship between them, and for more than 2 billion years the balance of life on Earth was held in the grip of these two co-existing forms. There are strong indications that life began to take a more robust approach through organised reproduction, and to acquire mobility. While fossil-hunters look to the start of the Cambrian, 541 million years ago, for their greatest trawl, and the start of the transformation overtaking life, there is increasing evidence from the Ediacaran, the 94 million years preceding the Cambrian, for a significant step-change in the evolution of life in what is known as the Avalon explosion.

This took place around 575 million years

ago and marks the earliest trace of complex multicellular organisms, right at the time Earth was thawing out from a major glaciation. This plays into the interpretation of evolution as having occurred at significant points when the greatest explosions of change took place, triggered perhaps by a subtle convergence of environmental factors that favoured a major niche-seeking track. Debate about reproduction in complex organisms during the Ediacaran took a new turn in mid-2015 with the publication of a scientific paper by a team from Cambridge University showing the earliest application of this asset, 565 million years ago.

Analytical techniques demonstrated that a distribution of rangemorphs (Ediacaran fossils), known as Fractofuss, appear to have been randomly distributed surrounded by 'daughter' products in clusters similar to groupings in modern plants. The communities appear to have grown through the older rangemorphs, seeding ejected water propagules, with the younger Fractofuss sending out reconnaissance parties to survey the new locations. Like plants today, rangemorphs were unable to move but could propagate through ejecting seeds, but it is not known if fertilisation was through sexual or asexual means. Scientists are beginning to realise that this mechanism enabled lateral fertilisation of new ground and possibly challenging environments, forcing adaptation but also determining productive new places in which to thrive. Life was on the move.

2.5 mm

An explosion of life

By the time of the Cambrian, a period enduring for 55.6 million years, oxygen levels in the atmosphere had reached 63% of where they are today, carbon dioxide levels were still very high – 16 times their pre-industrial level – while the temperature of Earth was 13°F (7°C) hotter than it is now and sea levels were rising across this period, from 98–295ft

ABOVE Brachiopods from the waters of the Devonian, 400 million years ago, when vascular plants began to appear on land and the transition to a terrestrial way of life began steadily to evolve. *(David Baker)*

LEFT The Carboniferous saw the explosion of tree and plant life on land, laying down the vast coal reserves exploited by industrialised man more than 300 million years later. Air breathing insects, giant spiders and gigantic arthropods crept upon the land, including the largest of them all, the Arthropleura, 28ft (8.5m) in length. *(I Sailko)*

Gingko biloba, thrived as seeding plants and took over, and conifers and ferns flourished. The first archosaurs appeared, related to birds and crocodiles, and precursors to the dinosaurs that would come to dominate the Jurassic.

When the Jurassic began 201 million years ago it saw the dawn of the age of reptiles and would see the emergence of the big dinosaurs, unrelated to reptiles and one of the most successful forms of life ever to walk the planet. Herbivorous sauropods grew to enormous size, some exceeding 100ft (30m) in length, while carnivorous saurischian dinosaurs hunted or scavenged on their less aggressive and poorly defended prey. Toward the end of the Jurassic the first feathered dinosaurs appeared and took to the air, where

they endure today as birds. With a completely non-reptilian bone and body structure, from the outset dinosaurs were made for flight, the saurischian types being light for their volume and ideally suited to the air.

The Jurassic environment changed dramatically over the Triassic, oxygen levels increasing to 130% of current values, carbon dioxide seven times the present level and with rising temperatures. These conditions were retained during the Cretaceous, which began 145 million years ago, with only a marginal reduction in CO_2 levels, but with oxygen levels reaching 150% of today's level. The Cretaceous was warm with high ocean levels and numerous shallow inland seas. The first leaf-bearing trees appeared along with flowering plants (angiosperms), and bees were prolific in co-existence and support for each other. Early mammals appeared in small numbers, and insects diversified into ants, termites and some precursors of butterflies, grasshoppers and wasps.

Rebirth

The great extinction at the end of the Cretaceous, 66 million years ago, marked a dramatic transformation in the fauna of Earth, wiping out in a few centuries a wide range of species as well as all the non-avian dinosaurs, leaving only the birds to endure to the present. During the Paleogene that followed, large hydrocarbon deposits in critical areas that today attract oil companies were laid down, notably in the Gulf of Mexico. On land, this was the age of mammals and the emergence of wide varieties of life with which we are very familiar today. Some mammals became large and stalked the wide and tree-filled spaces, others went up into the trees as primates.

The period since the great extinction at the end of the Cretaceous is divided into two epochs: the Tertiary, from 66 million years ago to 1.8 million years, and the Quaternary, to the present. Across Earth the environment fell back from the hot and humid conditions of the Mesozoic and entered the Cenozoic with a more stable and cyclical balance between glaciation and retreat, a completely different set of animals – most of them having emerged

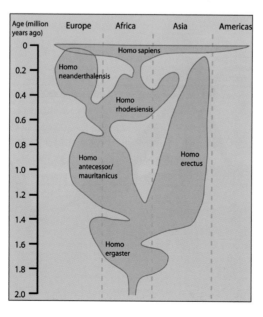

from the havoc of the great extinction – evolving and differentiating.

Proto-humans emerged within the last 5 million years and the several parallel species of hominid co-existed from around 4 million years ago, perhaps to less than 30,000 years ago. Ape-like Australopithecus, Homo habilis, Homo erectus – all preceded the arrival of modern humans, most probably out of Africa less than 150,000 years ago and for more than 100,000 years they would share space on Earth with Neanderthals, mixing their genes and merging habits and habitats, developing language, the written word, erecting great buildings, raising icons to their achievements, developing the ability to extract hydrocarbons and power their machines, take to the skies and travel to the Moon, building robots that are now departing the solar system.

But it is not only on the surface of Earth and above that life has demonstrated its powerful ability to diversify and adapt. Life is known to exist in some bizarre and demanding places, none perhaps more so than the reported sighting of flatfish and shrimp reportedly observed at the bottom of the Mariana Trench (see Chapter 3). The floor of the sea appeared to be smooth and undulating but devoid of recognisable life and others believe that what were believed to be fish swimming around were sea cucumber. However, in 2011 an expedition placed untethered landers on the sea floor and recorded giant single-cell amoebas more than 4in (10cm) in size of the class xenophyophores, which are frequently seen in large numbers on the sea floor and which often serve as hosts for other organisms.

In December 2014 the record for deep-sea fish was claimed when a new species was discovered at a depth of 26,724ft (8,145m) along with supergiant crustaceans. As humans explore further they are uncovering a fascinating world that has evolved over more than 4 billion years, and have the potential to reflect upon the awesome responsibility given to conscious beings who have the possibility of pondering and philosophising about their own existence, extrapolating goals and aspirations beyond their surroundings and living in coexistence with a changing biosphere. That alone endorses the ethic that co-existence is a survival tool rather than an opportunity to exploit irreplaceable resources.

Summary

- Life probably emerged before the Late Heavy Bombardment 3.9 billion years ago.
- Organic molecules are present throughout the universe.
- Prokaryotes and eukaryotes determined evolutionary tracks for life on Earth to follow.
- Life has proliferated in the last 541 million years, the most recent 12% of Earth's history.
- Great extinctions are a facet of Earth's history, clearing species out for renewal.

BELOW All animals have fought to maximise their opportunities and to exploit their environment but humans are the only known product of evolution to have systematically brought about the extinction of other species, such as the bison seen here wandering across an ice-age landscape, and not infrequently engaging in the intentional destruction of their own kind. *(David Baker)*

Chapter Eight

Future Earth

Earth is a machine that is running down as geological processes diminish, and the planet is slowly heating up due to the remorseless increase in solar output. But this is no perpetual motion machine. One day Earth will become uninhabitable and after that it will be absorbed by the red giant-in-waiting, sitting at the centre of our solar system, devoured by the star that gave it birth and spawned biological offspring.

OPPOSITE Slash and burn of brush and forest to clear land for farming began to pump carbon dioxide into the atmosphere more than 10,000 years ago, the start of an environmental catastrophe that mushroomed in the 20th century. In some locations these practices are still underway in the 21st century. *(David Baker)*

ABOVE Earth is no longer balanced by living organisms which secure niche areas for food and reproduction. The geophysical environment is now connected to sea and air in a coordinated exploitation by an ever burgeoning expansion of the human population, but growth which is destroying the very branch of the tree on which we sit. *(NASA)*

BELOW In its natural cycle the output of solar energy is increasing, but on a geological time span this has little short-term effect on the changing climate. Instead, solar energy is sought to free industry and domestic consumption from reliance on limited hydrocarbon fuels such as oil, gas and coal. These photo-voltaic cells near the German city of Hannover produce 1 megawatt of electrical power to the grid. *(David Baker)*

Much has been written about the future of Earth. Some opinion offers the view that humans should regard themselves as caretakers, as though humans have an ordained right to manage and look after it. On a local scale they do, for humans have declared themselves custodians of land and material resources, some of which were formed long before Earth itself began. Drawing upon Earth resources to provide energy and to power their cities and living spaces humans have for several centuries now used fossil fuels laid down across hundreds of millions of years.

Life has grown and proliferated on Earth without scavenging material resources other than those necessary for survival. The responsibility faced by humans is not for the custodial management of Earth but rather for the intelligent management of themselves as a species and not for controlling the planet, which would get along very nicely without them. If that is done effectively, humankind has a long and glorious future. If it is not, humans may not survive, for the scientific reasons that unfold in this chapter.

Because Earth is at present in the Holocene interglacial period of the Quaternary glaciation, under normal circumstances where no external factors were to change the cycles of ice ages we could anticipate a return to a new ice age within about 25,000 years. Because of the artificially increased levels of carbon dioxide and methane in the atmosphere, the onset of that new glacial period may be deferred to a more distant 75,000–100,000 years. But it is inevitable that it will return sooner or later.

If current levels of carbon dioxide and methane pumped into the atmosphere by human activity are reduced, and if hydrocarbons and other pollutants are constrained, and eventually eliminated within the next 200 years, the onset of the glaciation will be delayed by a mere 5,000 years or so. In the long run Earth will continue on its cyclical routine, and while humans may feel that the short-term consequences are catastrophic they will do little to significantly change the evolution of Earth in the longer term.

But there are other factors such as enhanced tectonic activity, possible significant changes to the climate by volcanic eruptions, in some cases predicted to be so violent in the next

200 years that they will transform the climate for centuries, and potentially catastrophic impact from asteroids, denizens of disaster and preferred Hollywood movies. Then there is the Moon, which is moving away from Earth at about 3.8cm (1.5in) per year and will eventually slow the rotation rate of Earth to a point where a 'day' lasts more than 50 hours and the Moon is locked on to one face of Earth.

The changeling

Earth as we know it may well survive the impact of human presence on the planet for the next several million years, although it is hard to imagine that humans can – as an organic addition to the multifarious forms of life on this remarkable world – escape the inevitable consequences of their actions. Unless humans evolve cerebrally to control their genetic predisposition to competitive conflict, material greed and the irresistible urge to expand populations disproportionate to the indigenous resources upon which all fellow humans rely, their fate is to go the way of all runaway life forms – extinction. Or worse, hanging on at the margins of subsistence, few in number, widely dispersed, dysfunctional and isolated, retreating to niched habitats where predators and prey are scarce, living on a planet where most life forms will have already been driven to extinction.

Alternatively, humans may choose to migrate from Earth in small numbers on the back of exploratory probing spacecraft across the solar system – first with robotic devices, later with people on expeditions to understand the rest of the solar system before establishing small colonies and isolated habitats. Only when a few people decide to remain, permanently attached to their new Eden, will migration become reality. But that is a very long way off, if, some say, inevitable as an antidote to extinction through resource depletion and unacceptable pollution and runaway ecological disasters.

Many of these matters are in the hands of humans themselves. Earth has been utterly transformed, and continues to be so, by the proliferation, variation and expansion of life in all its forms. It has always been so, for billions of years before humans came along. Which

is one reason why we cannot, as humans, claim prior place – we have no inherited right to unreasonably displace the rest of the natural world. We chase down its natural resources to exhaustion at our peril. But there are fundamental challenges to the *status quo*, on a planet that is in the grip of physical laws that will change everything for all living things, and us – eventually.

As the Sun moves past the middle phase of its stable life significant changes are going to affect Earth. To date the Sun has converted enough matter into energy to build 100 planets the size of Earth. At the rate of 4 million tonnes of matter converted into energy per second, there are major changes occurring right now at the centre of the Sun as fusion reactions convert hydrogen into helium producing neutrinos and solar radiation, gradually changing the nature of the Sun's interior.

End game

As helium atoms are formed they occupy less volume than hydrogen atoms that were fused to form them, and this displaces the Sun's core, causing it to shrink. Because of this the outer layers of the Sun move down closer to the core and this in turn exerts higher gravitational

ABOVE Oil production since the 1890s has fuelled the reciprocating engine for cars, boats, trains and aeroplanes, replacing the coal-fired steam engine of the preceding century. This in turn has created a demand linked to prosperity, accumulation of wealth and the shift to conspicuous consumption through manufactured goods produced and delivered by hydrocarbon engines. Carbon emissions from combusted hydrocarbons has created a heat-store in the atmosphere. *(David Baker)*

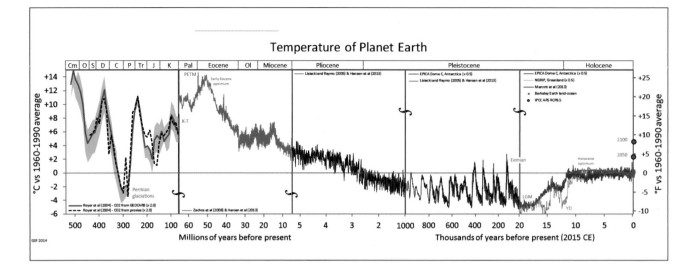

Temperature of Planet Earth

| Cm | O | S | D | C | P | Tr | J | K | Pal | Eocene | Ol | Miocene | Pliocene | Pleistocene | Holocene |

Royer et al (2004) - CO2 from GEOCARB (x 2.0)
Royer et al (2004) - CO2 from proxies (x 2.0)
Zachos et al (2008) & Hansen et al (2013)
Lisiecki and Raymo (2005) & Hansen et al (2013)
EPICA Dome C, Antarctica (x 0.5)
NGRIP, Greenland (x 0.5)
Marcott et al (2013)
Berkeley Earth land-ocean
IPCC AR5 RCP8.5

Millions of years before present

Thousands of years before present (2015 CE)

GSF 2014

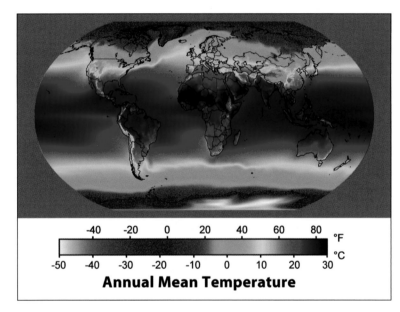

| -40 | -20 | 0 | 20 | 40 | 60 | 80 | °F |
| -50 | -40 | -30 | -20 | -10 | 0 | 10 | 20 | 30 | °C |

Annual Mean Temperature

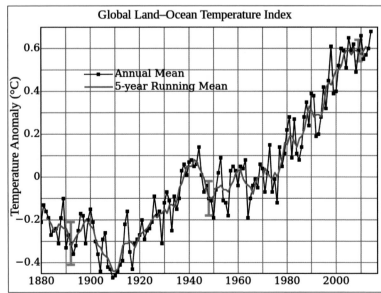

ABOVE Temperatures have generally stabilised over the last 500 million years, especially so since the Pliocene with fluctuations where glaciation has plunged to record lows. Since the Last Glacial Maximum temperatures have risen, only to begin a decline now reversed by human expansion within the last 10,000 years. *(NOAA)*

LEFT The average annual temperature distribution map shows heat generation in the Earth-engine. *(NOAA)*

forces which increase pressure, resisted only by an equal and reciprocal increase in the fusion rate to maintain thermal equilibrium. Over the almost 5-billion year life of the solar system to date, the star has increased its brightness by 30% and this rate is increasing so that right now it is growing at 1% every 100 million years.

After they become uncontrollably unstable, very large stars explode as supernova and shed vast quantities of their outer structure to interstellar space, bequeathing their manufactured materials to the stellar nurseries of the galaxy, where the next generation of stars will be born. But our Sun, as a G2 star, will not

LEFT In the 80 years since 1920, average temperatures increased by 1°C, an increase unprecedented in the geologic record, insofar as it is possible to say. An increase on this scale will inevitably result in significant changes to weather systems and eventually to the climate. *(NOAA)*

end its life as a supernova but will gradually expand and increase its surface area.

This sequence will itself last a very long time and, as Earth gradually warms under the increasing energy output from the Sun, life will have an increasingly unsuccessful struggle for survival. Complex and vulnerable endoskeletal life forms such as humans will be unable to live on the surface, although by then humans may have already left Earth for other, more comfortable places, in this solar system or beyond.

Any migration out from Earth will shift human habitation much farther out in the solar system, tracking the shifting ecosphere as it moves farther away from an increasingly hotter Sun, way out to the orbit of Saturn where its moon Titan may be the final resting place for humans prior to the demise of the Sun itself. This may transform the way humans evolve, for having severed the umbilical with mother Earth, the cradle of life within the solar system, it may be down to humankind to take with them other living things which would be unable to survive the planet's transformation.

About one billion years from now life as we know it will be impossible to sustain on the surface of Earth. At that time the Sun's energy will have increased by 10% compared to its current value and the temperature on the surface of Earth will have risen from its present value of 288 K (15°C) to 320 K (47°C). Over time, and by this date, the world's seas and oceans would have

subducted into the mantle until only 75% remains on the surface. But the increased temperatures will cause runaway evaporation, adding further heat to the atmosphere and raising its moisture level so that the stratosphere becomes saturated.

When that happens the water molecules would break down and solar ultraviolet light would induce photodissociation. This would cause hydrogen molecules to escape, further diminishing the ability of the oceans to replenish themselves, as the constituent molecules would break down and drift into space. Beyond that predictions are difficult but the extremes of temperature built up through the 'greenhouse effect' would see temperatures soar as high as 1,170 K (897°C). This is greatly in excess of the surface temperature of Venus, where lead runs like water in temperatures of 735 K (462°C).

As we have seen, there is several times as much water in the lower levels of the crust and the mantle today than there is at the surface, and for some time this would seep through, forming lakes and small inland seas. But the water in the upper atmosphere would rarely rain down, much of Earth's surface being arid and parched. Life would go into steep decline and habitats would not survive. Only humans could endure these conditions, and even then only in adapted habitats using artificial means to survive. The only redeeming feature of this scenario is that the changes would take place over millions of years – far longer than humans

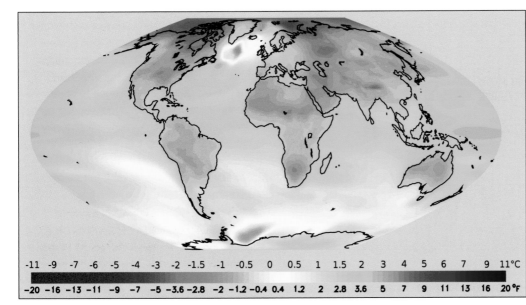

| -11 | -9 | -7 | -6 | -5 | -4 | -3 | -2 | -1.5 | -1 | -0.5 | 0 | 0.5 | 1 | 1.5 | 2 | 3 | 4 | 5 | 6 | 7 | 9 | 11°C |
| -20 | -16 | -13 | -11 | -9 | -7 | -5 | -3.6 | -2.8 | -2 | -1.2 | -0.4 | 0.4 | 1.2 | 2 | 2.8 | 3.6 | 5 | 7 | 9 | 11 | 13 | 16 | 20°F |

LEFT Global temperature averages lie about effects on a local scale. This map shows the level of temperature increase in specific locations which are inevitable between 1990 and 2050. Nothing can be done to reduce this because the forcing energy has already been released and will, over the next 35 years, have the effect shown here. *(NOAA)*

RIGHT Population growth is fuelling use of planetary resources although future trends are uncertain. However, much like temperature increases over the next several decades, the upward curve is inevitable for the foreseeable future, despite a reduction in birth rates among some Western countries. *(UN)*

BELOW Population growth rates for different geographic zones show the highest growth in Africa where cities are already outgrowing any other urban complex on Earth. Lagos, capital of Nigeria, already has a population of 13 million people, an extra 7 million within the outer suburbs. Within a couple of decades India will be the most populous country on Earth. *(UN)*

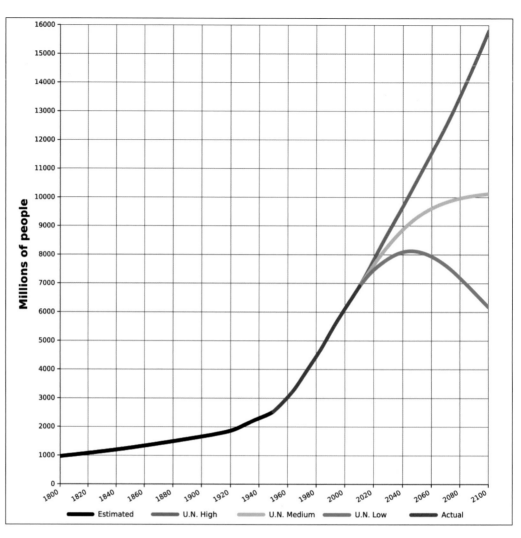

| Estimated | U.N. High | U.N. Medium | U.N. Low | Actual |

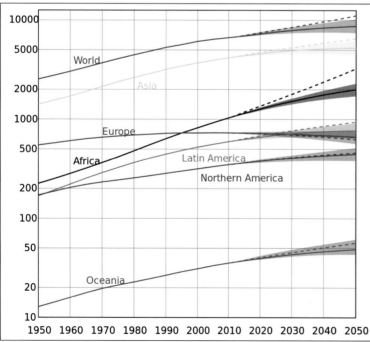

have been around so far in Earth's history, leaving adequate time to adapt.

But for people, that future is beyond prediction and entirely dependent on whether humans manage to survive the next several million years. If they do, it is likely that human populations will have spread across the solar system and that the first voyages to other star systems will have already taken place. It is not so far-fetched. The technology exists today to set sail for the stars. It is an engineering challenge alone, and not a possibility constrained by physical laws. It merely requires the commitment – or, as so much has in the past record of human endeavour, the need to survive and to thrive. For as humans, we are restless and not content to sit still and await our fate.

Although it is unimaginably far off in the future, Earth will one day succumb to the evolutionary life of its parent star – the Sun, which will, by about 1.4 billion years in the future, expand into a red giant. At that time its outer surface will lie at or beyond the orbit of Earth and its intense heat will consume the very planet to which it had bequeathed all of its life-supporting properties.

Earth will remain in its present position because the overall mass of the Sun will not have changed significantly, relative to what it is today. Therefore, the Sun's gravitational force will not have changed to dramatically alter Earth's orbit. But the outer envelope of the Sun's surface could induce drag on the planet and this may drive an increase in orbital eccentricity and the potential for Earth to spiral into the Sun itself.

We saw at the beginning of this book how the Sun started out as a main-sequence star with a stable life of around 9 billion years. Because the solar system is less than 5 billion years old there is still a lot of time left for humans to sort out their differences (if there are any humans left that far in the future!). But, as we can now see, we may not have as much time as we thought.

Space travel has given us the technical and scientific means to measure, with extraordinary precision, precisely where the various planets are and to characterise with equal precision the

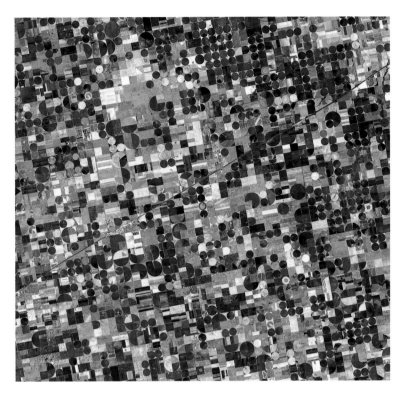

orbital paths they follow. As we have already seen, planets are in unstable and elliptical orbits that change over time. Some scientists have carefully calculated that the orbit of the inner planet Mercury could grow even more eccentric so that a close pass of Venus might propel it either out of the solar system or on to a collision course with Earth.

ABOVE The need to feed an ever expanding population is driving farming to industrialise production, evidenced by these circular crop fields in Texas showing healthy corn and sorghum in green and wheat in red. *(David Baker)*

LEFT Miles of rice terraces in Yunnan Province, China, linking historic means of food production with the modern world. The Asian diet is becoming more reliant on meat products which require much greater draw on planetary resources due to the lengthy cycle of several years to deliver meat to the plate rather than harvested cereal crops delivered to the table within a single season of sowing. *(David Baker)*

ABOVE The transition for humans from hunter-gatherer to famer meant switching from nomadic groups to settled communities but the desire for choice meat cuts drives up resource demand to produce high-quality beef, typified by this cattle farm in Hyogo Prefecture producing Kobe beef. (David Baker)

Uncomfortably for us, just as the Jupiter–Saturn system cleared the inner region of the solar system in the early days of its evolution, so might Earth and Venus exert gravitational attraction over the lesser planets. Such a potential Earth-collision course is also just possible for the orbit of Mars, which is already eccentrically biased toward a potential Earth collision path. But neither of these events is likely to occur within the next 3 billion years, by which time, as we have determined, humans will probably not need Earth for their survival.

Less predictable, however, are events associated with another event which is more soundly based on accurate measurement and physical laws applicable, as far as we can tell,

everywhere in the Universe. It is a fate that is most surely going to happen and, like it or not, one we cannot avoid.

Game over?

Within 2.4 billion years Earth's orbit, and those of every other body in the solar system – including perhaps the Sun itself – will be affected to varying degrees by the impact of the Andromeda galaxy as it ploughs into the Milky Way. Because the solar system lies approximately two thirds of the way from the centre to the outer edge of this spiral galaxy, the Sun's position within the Milky Way will make it potentially vulnerable.

As the largest member of the local group of 54 galaxies, most of them small, Andromeda is the only galaxy that can be seen with the naked eye. The local group is itself part of a super-cluster 110 million light years across containing around 100 galaxy groups and clusters. Most of these are generally increasing their distance from each other, but, down at the (relatively) local level, Andromeda is rushing toward us at a speed of 120km/sec f(74.5 miles/sec, or 268,200mph). Andromeda has a diameter of 220,000 light years, sizeably larger than the Milky Way, which is about 180,000 light years across.

When these two cosmic leviathans collide the greater number of stars will simply pass through space between each other, hardly

RIGHT The drive to harvest the seas is forced by expanding populations and a demand for high quality fish. Challenged by difficult conditions, the French fishing vessel, Alf, struggles to get a catch in the Irish Sea. (MoD)

aware of the encounter. There is just *so* much space out there that the distances between stars is enormous. But at the central region of each galaxy, where stellar populations are dense, encounters will be more likely and the magnetic fields, plasma, gas and dust associated with each island universe will interact, which could seriously perturb stars within local groups, slowly moving them away from their present orbital positions in the galaxy and, just conceivably, ejecting Earth from its path around the Sun.

These events are conjectural, but scientists have calculated that as the long and extended galactic tails intertwine there is a better than one in ten chance that the solar system will be pulled into the tidal tail of the Milky Way. They also believe there is a 3% chance that our solar system will be transferred to the outer arms of the unwelcome galactic intruder, permanently relocating the Sun and it's planets to Andromeda. Added to this is the collision-enhanced probability of other stars passing sufficiently close as to move the solar system with respect to its orbital path in the galaxy.

If Earth does avoid being struck by a deflected Mercury or Mars, which is far more likely than not, and if it is not transferred to another galaxy through converging collision, which is improbable at least, our solar system will certainly go through a dramatic change about 5.4 billion years from now.

As the outer envelope of the Sun burns

hydrogen in the shell surrounding the core during the red giant phase, the Sun will increase its luminosity by a factor of 2,700, cooling down its surface to a mere 2,600 K. But this is beyond our concern, for by this time Earth may well have been consumed, since about 7 billion years from now the Sun will enter the helium-burning phase, where it loses almost a third of its former mass and ejects shells of matter into interstellar space as planetary nebula.

This phase fuses helium to carbon and oxygen at its core through what is called the triple-alpha process, where the core temperature of the Sun is 10^8 K and helium nuclei fuse faster than their off-product, beryllium-8, which decays back down

ABOVE In a single catch, 400 tonnes of mackerel is caught by a Chilean net system. Overfishing in many areas of the world is destabilising the food chain and driving some predators to extinction while opening colossal growth in predated species now free to multiply. *(David Baker)*

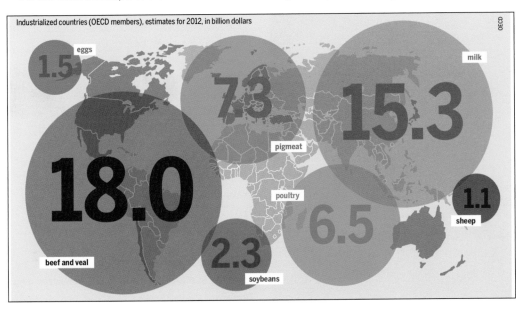

Industrialized countries (OECD members), estimates for 2012, in billion dollars

OECD

eggs **1.5**

7.3 pigmeat

milk **15.3**

18.0

poultry

6.5 sheep **1.1**

beef and veal

2.3 soybeans

LEFT The demand for food is driving up subsidies (shown here) due to an imbalance between supply and demand, forcing mechanised farming and industrial-style production. Government subsidies are already draining government funds away from financial needs for combating poverty and destitution among war-torn territories. *(OECD)*

into two helium nuclei. As the beryllium-8 production rate outpaces the decay rate its atoms fuse with helium nuclei to form carbon-12, and so on.

By this time Earth will be no more and the star which once gave life to an evolving Earth, every bit a living organism as vital as biological life itself, will have collapsed down to a white dwarf composed of degenerate matter in a volume no bigger than Earth was when it existed. White dwarfs are charred and blackened remnants with 70% of the mass of the star remaining. They are dense, solid bodies held up solely by electron valency shells, lacking the mass necessary to crush them into neutron stars or black holes.

The Sun will have cooled to a mere 5 K and remaining planets such as the gaseous outer giants will be picked off one by one as passing stars exert gravitational attraction on them. The white dwarf remnant will exist for quadrillions of years. Long before then, as the universe expands and accelerates, the dispersal of matter in the galaxies will leave no opportunity for new stars to form. Quite literally, across the universe the lights will go out and nothing will be left except the manufactured materials from long dead stars and empty space.

As expansion of space itself across the broad swathes of the universe accelerates on the macro-scale, at the galaxy level all star-making materials will have been consumed and they will grow dimmer until starlight is all but gone, except for the very small stars – the last remnants of a glorious age of star-making and galaxy-building, of planetary formation and of burgeoning life.

RIGHT Some people believe the only solution for a planet stressed by an overwhelming exploitation of its resources is to begin colonisation of the solar system, such as transforming Mars into a habitable environment, a process known as "terraforming", represented here in an artist's depiction. (David Baker)

LEFT Earth's fate one day is to become a barren and lifeless world with boiling oceans, as depicted in this impression of an ever-expanding Sun and its effect upon a bleached planet. (NASA)

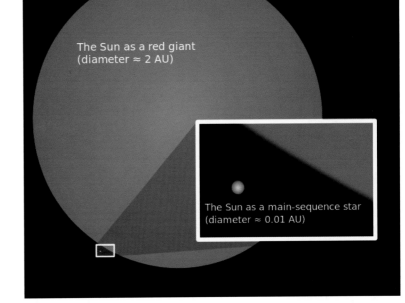

The Sun as a red giant
(diameter ≈ 2 AU)

The Sun as a main-sequence star
(diameter ≈ 0.01 AU)

Where once the most magnificent object in the solar system, our Sun, shone forth into the blackness of the void, giving life and growing sentient beings on a nearby rock, the very product of that wonder will have gone, immortalised perhaps by humans and other living things which have made new worlds of their own from the product of a universe that will have dispersed its material fabric and left nothing of its former glory.

Life that began on Earth and grew sentient beings into galactic explorers may by then have produced humans who will create their own worlds, giant artificial planets assembled from the debris of a dying universe, building a new and unimaginably different place in which to exist.

That is a future we can hardly begin to imagine. Perhaps, if we do survive, the legacy of one small planet – Earth – where it all began, less than 5 billion years ago, will be for life to ultimately prevail, perhaps beyond material existence, and to become part of the very fabric of space and time itself. Earth does not need humans, but for the foreseeable future we need Earth. We are not custodians of the planet, for, given our feeble abilities to control events beyond our own limited horizons, that is too arrogant a position to assume. But our own future on Earth depends upon the decisions we make today and tomorrow; because the most vulnerable element is not Earth at all, but our own future and our own survival. We are the ones who need saving, not the planet. On its own, it has been successfully accommodating change and diversity for several billions of years already.

Summary

- Life on Earth is restricted to the quiet portion of the Sun's stable life.
- As the Sun increases its energy output, land will become arid and oceans on Earth will boil.
- As it expands, the Sun will eventually consume Earth and destroy remaining life forms.
- In a distant future Earth may be captured by a colliding galaxy – Andromeda.

ABOVE When it expands to the Red Giant phase, the outer envelope of the Sun will brush Earth and create an inferno of molten rock, returning the planet to the form it had when coming together at the birth of the solar system. *(NASA)*

RIGHT For millions of years to come Earth will remain a potential haven for life. However, it cannot continue to support an ever-expanding human population demanding increasingly affluent lifestyles exploiting limited resources that will, inevitably, run dry. Human achievement has been extraordinary, but we are now the sole hominid species remaining. How we organise the next several centuries could decide the fate of the entire human family. For its part, Earth will still be here. *(NASA)*

Index